MW00476177

PHARMACEUTICAL PRODUCT BRANDING STRATEGIES

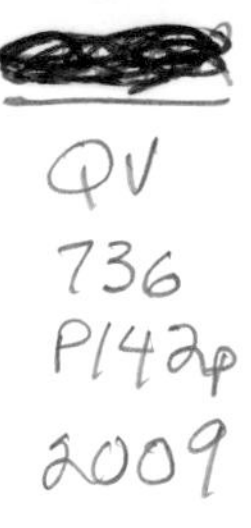

PHARMACEUTICAL PRODUCT BRANDING STRATEGIES

Simulating Patient Flow and Portfolio Dynamics

Second Edition

Mark Paich, Ph.D.
Lexidyne, LLC
Colorado Springs, Colorado, USA

Corey Peck
Lexidyne, LLC
Colorado Springs, Colorado, USA

Jason Valant
Lexidyne, LLC
Colorado Springs, Colorado, USA

informa
healthcare

New York London

Informa Healthcare USA, Inc.
52 Vanderbilt Avenue
New York, NY 10017

Printed in the United States of America on acid-free paper
10 9 8 7 6 5 4 3 2 1

International Standard Book Number-10: 1-4200-8770-3 (Hardcover)
International Standard Book Number-13: 978-1-4200-8770-3 (Hardcover)

Library of Congress Cataloging-in-Publication Data

Paich, Mark.
Pharmaceutical product branding strategies : simulating patient flow and portfolio dynamics / Mark Paich, Corey Peck, Jason Valant. – 2nd ed.
p. ; cm.
Rev. ed. of: Pharmaceutical product strategy. c2005.
Includes bibliographical references and index.
ISBN-13: 978-1-4200-8770-3 (hardcover : alk. paper)
ISBN-10: 1-4200-8770-3 (hardcover : alk. paper) 1. Drugs–Marketing.
2. Brand name products–Mathematical models. 3. Pharmaceutical industry–Mathematical models. I. Peck, Corey. II. Valant, Jason. III. Paich, Mark. Pharmaceutical product strategy. IV. Title.
[DNLM: 1. Drug Industry–economics. 2. Marketing–methods. 3. Models, Economic. QV 736 P142p 2009]
HD9665.5.P35 2009
615′.10688–dc22

2008048873

For Corporate Sales and Reprint Permission call 212-520-2700 or write to: Sales Department, 52 Vanderbilt Avenue, 16th floor, New York, NY 10017.

Visit the Informa Web site at
www.informa.com

and the Informa Healthcare Web site at
www.informahealthcare.com

Preface

As we put the finishing touches on the second edition of this book, it is truly overwhelming to think back on the journey that has brought us to this point. Our initial forays into using the methods outlined in the pages, which follow began almost 20 years ago, at the time when this framework was both relatively obscure and not readily accepted within the pharmaceutical sector. The past two decades have seen huge changes in the willingness of industry participants to embrace the Dynamic Modeling methodology. In part, this reflects the increased prominence of "data detectives"—as popularized in best-selling books such as *Freakonomics* and *Super Crunchers*. We believe one component of the story comes from sheer necessity—increased market pressures have forced pharmaceutical and biotech companies alike to seek out and embrace better frameworks for raw analytics and judgment-based elements of the Brand Planning process. And we like to think that our continued efforts to explain, advocate, and practice the techniques outlined in this book have increased both the breadth and depth of penetration of this methodology within the pharmaceutical industry.

This book was written for nontechnical readers, although the "guts" of various simulation engines and analytical approaches are in fact fairly intense mathematically. We have found that the concepts at the heart of the Dynamic Modeling paradigm are readily accessible, however, if they are presented in the right context. Readers will be the final judges as to whether we have achieved that mission. In order to compare our framework with existing methodologies, this book progresses from an evaluation of traditional paradigms employed by most brand planning teams to the more advanced and insightful picture offered by Dynamic Modeling. Specifically, the chapters are organized as follows:

- Chapter 1 details the traditional approach to Brand Planning in the pharmaceutical industry and highlights some limitations to existing approaches.
- Chapter 2 introduces the Dynamic Modeling framework, presents its key advantages over current approaches, and offers examples of important insights this methodology has provided to some of our clients.
- Chapter 3 guides the reader through the Dynamic Modeling language that is used in subsequent examples.
- Chapter 4 depicts an important sector of this approach called Patient Flow, which illustrates how potential, current, and former patients move through the marketplace. A simplified example that includes epidemiology and treatment algorithms is introduced and extended in subsequent chapters.

- Chapter 5 shows how the example from Chapter 4, which we call the Standard Template, can be extended to meet specific conditions seen in certain pharmaceutical markets.
- Chapter 6 describes data formats, sources, and analysis that are important in populating the Patient Flow sector of the Standard Template.
- Chapter 7 employs the Dynamic Modeling language to address situations (involving patient dynamics) that do not easily fit the Standard Template.
- Chapter 8 introduces the physician component of pharmaceutical marketplaces by showing the structure and analysis surrounding Doctor Adoption of prescription therapies.
- Chapter 9 introduces a sophisticated new technique increasingly incorporated into the Dynamic Modeling approach called Agent-Based Modeling.
- Chapter 10 details how therapies can be evaluated on dimensions of Treatment Attractiveness and how such utility calculations impact marketplace dynamics.
- Chapter 11 joins the three main sectors of the Standard Template to show how pharmaceutical markets are driven by the dynamic interaction of patient flow, Doctor Adoption, and Treatment Attractiveness.
- Chapter 12 illustrates how fully integrated models can be used to better understand market dynamics and develop more effective marketing strategies.
- Chapter 13 details how the Dynamic Modeling approach can be used to investigate network effects in disease markets and quantify any first mover advantage in reaching the marketplace.
- Chapter 14 addresses how uncertainty in model inputs can be handled, particularly in the cases of latent markets in which no historical data is available.
- Chapter 15 shows how individual models of specific indications can be combined to create an overall portfolio model.
- Chapter 16 demonstrates how the Dynamic Modeling approach can be used to analyze pipeline/portfolio dynamics associated with the drug development process.

Much of this work uses a progressively more detailed model of a hypothetic pharmaceutical market to illustrate the methodology and resulting impact on brand planning. This model is intended to be illustrative, not definitive, and the interested reader will certainly be able to project how such a basic framework can be augmented and extended.

We hope you will enjoy learning about the Dynamic Modeling process as much as we have enjoyed practicing and writing about it.

Mark Paich
Corey Peck
Jason Valant

Acknowledgments

This work would not have been possible without the dedicated efforts of a widely dispersed team—the members of which go far beyond the names listed on the front cover. We would like to thank Tom Waters, Harvey Kane, Steve Zollo, and the rest of the talented folks at CRC Press for their efforts in getting the first edition of this book to press. More recently, Michelle Schmitt-DeBonis and Sherri Niziolek at Informa Healthcare gave us the opportunity to revise and update this second edition and guide us to its successful completion. Without their direction and support, the process would never have gotten off the ground.

Our thanks go to our colleagues, clients, and business associates who generously offered their insights and contributions to this work, which greatly benefits from the real-world insights of Cynthia Willey, Sue Steven, Steve Peterson, Julie Stafford, and Steven Bloom and the vignettes they contributed. A very special thanks goes to Julie Stafford at AXIA Management Consultants—the book is much better for her positioning help and for providing thoughtful and timely feedback throughout the development process. Lindsey Carmichael, Kirk Solo, and Steve Peterson were instrumental in giving content suggestions and directional feedback during both the early and late stages of this book's development. Molly Shea was a valuable behind-the-scenes member of our team in terms of industry research, chapter editing, and content discussion.

Major events in life sometimes result from a series of random coincidences. The idea for this book came in the summer of 2002 when our flight was cancelled and we found ourselves stranded at the Philadelphia airport. The hotel shuttle service was 30 minutes late to arrive, and we passed the time by beginning to formulate the content and scope of this work. Had the flight departed as planned or had the shuttle been on schedule, it is unlikely we would have embarked on this publishing journey. So in truth, we are grateful for flight cancellations and we owe a huge debt of thanks to van delay.

As practitioners of the Dynamic Modeling approach, we are blessed to benefit from a unique vantage point created by those who have come before us. Most notably our heartfelt appreciation goes to the irreplaceable Barry Richmond—you are sorely missed.

To our families and friends who endured, with lasting patience, not only the development of this book but the numerous consulting engagements (usually far from home) on which it is based; we can only give our everlasting thanks for your love and understanding.

To our parents who so selflessly gave each of us the support and resulting freedom to pursue whatever paths life's journey presented to us. We are forever in your debt.

Much has happened since the publication of the first edition of this book in 2004. Most notably, we have seen an expansion of both our professional and personal families and it is our hope that this work is enjoyed by many important people from both of those circles.

Contents

1

Traditional Approaches to Brand Planning: The Disconnect Between Product Strategies and Commercial Evaluations

INTRODUCTION

This chapter will cover the following topics:

- Challenges facing the pharmaceutical industry
- Inadequate analytic methodologies in the pharmaceutical world
- Existing approach to strategic/Brand Planning
- Methodologies for commercial assessment
- Lack of integration between brand positioning and commercial assessment
- Risks associated with ineffective decision making

The American public has a love–hate relationship with pharmaceutical companies. On the one hand, we are avid consumers of the products they develop and distribute. More than 3.8 billion prescriptions were dispensed in the United States in 2007 (1), representing more than 10 "scripts" per year for each and every citizen in the country. According to one study, an astonishing 74% of the population is using at least one prescription medication at any given time (2). The aging of the population, the increase in conditions such as diabetes and obesity, and the creation of government programs such as Medicare Part D are all likely to further increase the dependence of Americans on pharmaceutical therapies.

Despite such a clear and widespread prescription usage, Americans continue to maintain an unfavorable perception of the pharmaceutical industry as a whole. According to Datamonitor, in 2005, less than 13% of consumers feel that the information provided by industry firms is more trustworthy than that provided by other organizations. Complaints on the part of consumers are on the rise and the resulting political pressures helped push through the Food and Drug Administration

Amendments Act of 2007—a sweeping reform bill designed to address some major industry problems. And the election cycle of 2008 saw a number of candidates bashing "Big Pharma" (along with other targets such as "Big Oil" and "Big Insurance") as one of the main reasons for skyrocketing health care costs.

The pharmaceutical industry is well aware of the conflicting opinions on the part of its customer base, but is facing internal challenges of its own. Many firms are attempting to change in response to evolving marketplace conditions that threaten the "blockbuster drug" business model of the 1990s. New compounds are increasingly difficult to discover and develop, and the regulatory approval process reflects this tightening of developmental pipelines. During 2005 to 2007, the Food and Drug Administration (FDA) approved only half the number of new compounds as it had only a decade before (3). And fewer than 10% of these newly approved compounds are expected to ultimately generate sales of even $350 million annually (4)—far below the $1 billion annual revenue benchmark usually reserved for "blockbuster" status. Five different blockbuster drugs went off-patent in 2006 and more such transitions loom large on the horizon; Lipitor (the world's biggest selling pharmaceutical product) has only a few years of patent protection remaining (5). Price pressures, both from the public and private sectors, have made headlines nationwide—witness the plan in 2004 by the state of Illinois to import pharmaceuticals from Canada in order to cut costs. These situations have made the pharmaceutical industry as a whole seem vulnerable in the face of new challenges, new realities regarding drug development, new competition from biotechnology and the emerging world of genomics, and new expectations on the part of consumers and managed care providers.

The gloomy picture for the future of traditional pharmaceuticals, however, is being made brighter due to an emerging environment of new approaches to both analyze disease marketplaces and develop effective marketing strategies within them. Acceptance of new analytical techniques, associated computing power, and a wide variety of patient and physician data constitute a "perfect storm" of conditions to support a new paradigm for pharmaceutical marketers and executives alike. The need for a new approach has never been greater, and the results can often have huge impacts regarding the strategic decisions designed to create and maintain competitive advantage to the pharmaceutical industry.

These key business decisions have the potential to be supported by the wide availability of data relevant to pharmaceutical businesses, if only a coherent and consistent framework can be utilized to incorporate and leverage it. We like to tell our clients that *data ≠ information ≠ knowledge ≠ wisdom*. Only through careful examination of data, using cutting-edge analytical and simulation techniques, can the true power of the vast amount of industry-wide data be truly unleashed.

Extensive records regarding patient and physician behavior are now readily available from claims databases and third-party data providers, which in longitudinal form make it possible to track the dynamics of various patient segments and physician prescribing patterns over time. Extensive competitive intelligence makes product pipelines of industry players transparent to any analyst with an Internet connection. Masses of raw data are useless without an interpretive framework that synthesizes it into information that can generate learning and create the foundation for significant and sustained competitive advantage. And in the ever-changing pharmaceutical world, the Dynamic Modeling framework is a highly effective organizing tool to capture, analyze, and ultimately leverage the complex dynamics driving industry behavior.

In our consulting practice over the last 15 years, we have assisted our clients in advancing their Brand Planning approaches and associated evaluations of commercial potential for **New Chemical/Molecular Entities (NCEs or NMEs)** or **New Biologic Entities (NBEs)** toward sophisticated methodologies that have been popular in other industries for years. Dynamic Modeling represents a cutting-edge approach for operationally defining the set of interrelationships that drive the behavior of pharmaceutical marketplaces.* This approach is based on established techniques from the world of operations research and system dynamics, and adds value in at least two distinct ways.

1. The Dynamic Modeling approach creates a means to test the linkages between cause and effect in various pharmaceutical settings. Such simulation-based models can be used to operationally define the expected outcomes of a set of strategic decisions, resulting in better strategic plans and a more complete understanding of the sets of key relationships governing them.
2. Dynamic models provide a useful operational input to the forecasting process by establishing a framework to integrate the knowledge bases that exist within a pharmaceutical firm's functional divisions. The approach establishes practical boundaries on the realm of possible outcomes from a commercial assessment, providing a solid foundation for reasonability checks of the commercial potential for an NCE. As such, dynamic models provide another lens through which the future behavior of a disease marketplace, and the position of a particular compound within it, can be seen.

Before investigating the details regarding the Dynamic Modeling framework, however, it is useful to review the typical process for developing Brand Plans for an NCE, as well as highlight some deficiencies in the existing approaches of strategy development in the pharmaceutical world.

TYPICAL APPROACH FOR STRATEGIC/BRAND PLANNING IN PHARMACEUTICAL INDUSTRY

The pharmaceutical industry has been a large part of our consulting practice over the last decade, and many of our engagements are headed by marketers in charge of a particular developmental NCE.† These marketers often have the title of Brand Manager for a particular compound and are typically charged with developing a

* Operational representations explicitly detail how processes occur in the real world rather than relying on statistical correlations or imprecise understanding of component relationships (see chap. 2).

† Dynamic Modeling is not limited to new product introductions and can also be used to analyze compounds already on the market (see chaps. 4 and 5).

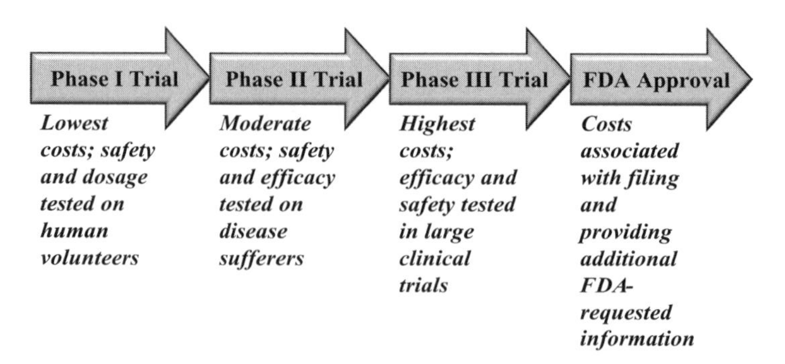

Figure 1 Clinical trial and FDA approval process for developmental NCE.

strategic plan for the NCE in late Phase II or early Phase III trials on its way to the approval by the Food and Drug Administration (FDA), as seen in Figure 1.

Results from the relatively inexpensive Phase I clinical trials force pharmaceutical firms to make critical decisions regarding the future of the NCE and the associated resources allocated to it. Sometime after the beginning with Phase II but often not until Phase III, senior executives will commission a Brand Plan to assess the technical and commercial potential of the new compound.* Brand Managers typically assemble a cross-functional team comprised of representatives from the firm's marketing, clinical/medical, health outcomes, and forecasting departments† to evaluate the financial potential of the NCE. If the new compound is currently in Phase II, the evaluation process might also help decide on a testing plan for the subsequent clinical trials. For example, the strengths and weaknesses of currently available products often suggest a Phase III clinical trial design to demonstrate the superiority of the developmental drug.

The Brand Plan team is assembled to guide the compound through the development process toward the launch of the product—a complicated endeavor involving numerous milestones and a large investment of human/financial resources. However, from a marketing perspective a Brand Plan has two key objectives.

Key Components of a Brand Plan for an NCE

1. A set of product strategies for the short-term time horizon (1–3 years)
2. An early commercial assessment of the compound's financial potential

Pharmaceutical firms over the last decade have increasingly adopted the practice of cross-functional coordination in order to leverage various types of knowledge within the organization, and marketers often sit at a conceptual hub of company functions when it comes to early assessment of NCE market potential.

* Recognizing, of course, the possibility of compound failure during the development process (see chap. 16).

† Representatives from regulatory, legal, and even operations might be a part of the Brand Plan team as well, but are not included in this example for the sake of simplicity.

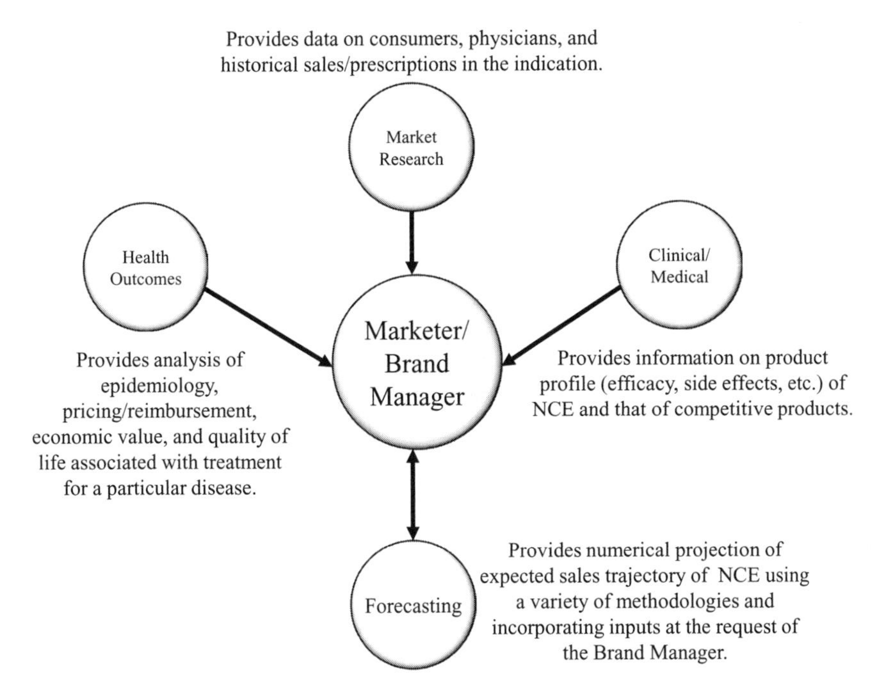

Figure 2 Interaction of various functional areas in determining a Brand Plan.

In order to develop an effective Brand Plan, the Marketer/Brand Manager is often the "gravitational force" behind such cross-functional interactions, as depicted in Figure 2.

Clinical/Medical

Often physicians who previously practiced medicine in a specific disease area, Clinical/Medical experts are knowledgeable in the pros and cons of various currently available treatments and have the technical expertise to understand and evaluate the product profile of the NCE, especially as relates to understanding the design and results of clinical trials. Their knowledge of standard practice and corresponding treatment paradigms provides excellent background into how patients are viewed and subsequently treated by physicians. Clinical/Medical experts help place the NCE within a competitive market framework, often by working with the Marketer/Brand Manager to develop a Strengths, Weaknesses, Opportunities, and Threats (SWOT) analysis.

Health Outcomes

In order to assess the impact of a disease on individuals or the economy, Health Outcomes resources analyze marketplace factors such as epidemiology, the economic benefit of various forms of treatment, pricing and reimbursement, and lifestyle measures of the effect of a disease on quality of life metrics. Epidemiology is an increasingly important part of this analysis, and is based on evolving population demographics and changing prevalence, diagnosis, and treatment rates. Health outcomes personnel may provide information how the disease affects specific patient segments at the request of the Marketer/Brand Manager, but their primary contribution is the overall assessment of how a particular disease impacts the health of those suffering from it and lives of those around them.

Market Research

Market Researchers represent the "voice" of the marketplace and specialize in collecting, analyzing, and communicating information on the indication marketplace and individual existing treatment options within it. Market Researchers design primary market research studies to collect data on physician/consumer reactions to NCE product profile, pricing, packaging, etc., often using a conjoint analysis framework. They may also query secondary data sources to obtain specific patient-level data or information on physician behavior. Marketers/Brand Managers often request specific information from Market Researchers in order to better understand important facts related to a disease marketplace.

Forecasting

As experts in pharmaceutical marketplace analytics and projection methodologies, Forecasters can assess the commercial potential of the developmental compound and work closely with the Marketer/Brand Manager to develop a prescription and/or revenue projection for the NCE. Forecasters are often technically oriented and their role is to provide a numerical forecast for the compound, not to focus on the strategic implications of their analysis. However, their interaction with the other members of the Brand Plan team often clarifies the assumptions associated with the disease marketplace and the place of the NCE within it.

Strategic Outcomes from the Brand Planning Process

The efforts of the Brand Plan team result in a comprehensive look at the disease marketplace, the competitive landscape, the pros and cons of various currently available and pipeline treatment options, the assessment of the unmet medical needs in the market, and other information designed to inform decision makers about the conditions into which the new compound might be introduced. From a marketing standpoint, however, the key outcome of the Brand Planning process is the concept of brand positioning, i.e., how the compound will be positioned with respect to its target audience, its expected benefits, its key reasons for trial/usage, etc. Brand positioning helps establishing a series of product strategies created to leverage the collective knowledge of the disease market and effectively use resources to increase the uptake of the NCE.* The strategies are often categorized

* In our consulting practice, we refer to these issues as affecting the Place and Pace for a compound (see chap. 12).

by areas of target influence, such as patient and/or physician segmentation, impact on the regulatory environment, effect on pricing/reimbursement, publication strategy, etc. In total, these strategic initiatives are designed to meet some key objectives regarding the performance of the NCE in the marketplace—goals by which the success of the launch of the compound will be measured. The strategies resulting from the Brand Planning process are then rolled out to the operations personnel in various markets in order to determine the tactical approach to support them.

Commercial Evaluations from the Brand Planning Process

At least in part due to the importance of a sound commercial assessment of the NCE, forecasting is usually an interactive and iterative process in which information gleaned from Market Research, Health Outcomes, and Clinical/Medical is given due consideration in developing a financial projection for the NCE. Forecasts in the pharmaceutical world rarely rely on one methodology or a single set of input assumptions. They may be derived from volume analysis, extrapolated from market financial projections, or even calculated based on patient algorithms. Analogs of other drugs in an indication marketplace, statistical formulations, econometric models, and patient-based calculations may all be employed and compared as ways to evaluate the commercial potential of an NCE.

In our experience in the pharmaceutical world, however, these types of forecasting analyses boil down to two sets of aggregate assumptions about an indication marketplace and the position of the NCE within it, as shown in Figure 3.

In Figure 3, the Expected Indication Prescriptions/Sales is usually calculated from an extrapolation of historical data, perhaps modified by epidemiologic trends or expectations of advances in the efficacy of various treatment options. Expected Share of Prescriptions/Sales by NCE is generally derived from analysis of product analogs, comparison of the NCE to existing treatment offerings in the marketplace, and team judgment regarding the level and type of marketing support for the compound.

Forecasted values for market size and NCE share are then multiplied together to arrive at an **expected sales trajectory** for the compound (Expected NCE Prescriptions/Sales in Figure 3). Note that this projection includes three key concepts:

1. The peak volume/sales figure.
2. The speed at which the NCE will reach this maximum revenue potential.
3. The shape of the falloff from peak sales.

The estimates of an NCE's expected sales trajectory can be analyzed using a Monte Carlo technique (see chaps. 14 and 16) that produces a range of possible outcomes, given the uncertainty about the actual magnitudes of various input assumptions. The results of a Monte Carlo analysis are often expressed as Stretch, Most Likely, and Worst case scenarios regarding the commercial assessment of the compound, as shown in Figure 4.

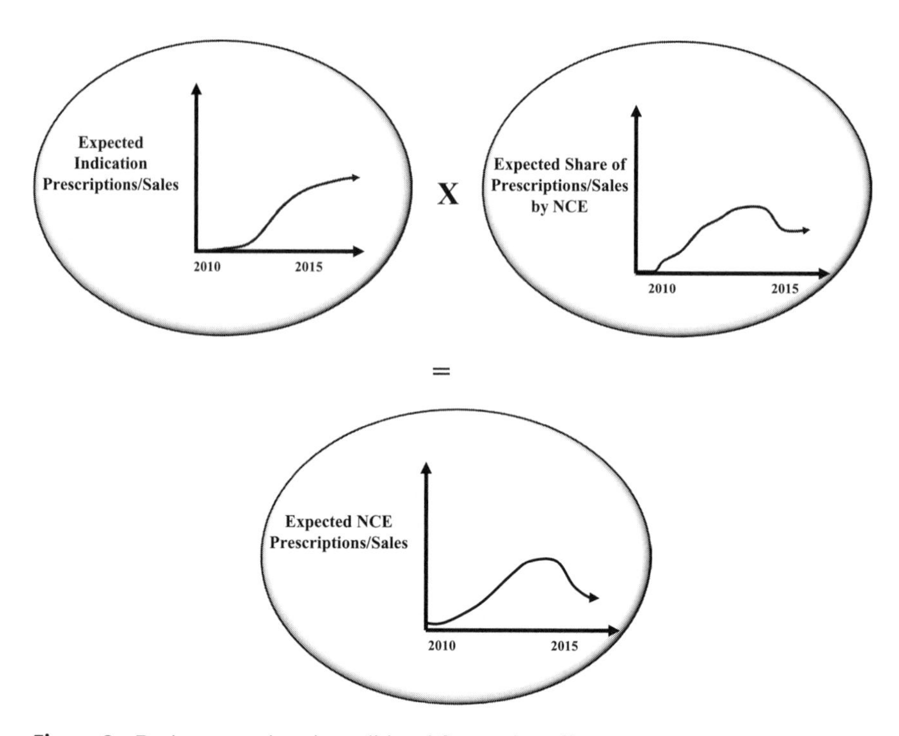

Figure 3 Basic assumptions in traditional forecasting efforts.

The result of the typical forecasting process becomes an important part of the Brand Plan for the NCE, and this forecast is usually presented as the chain of command within a pharmaceutical firm. The early commercial assessment of the NCE often dictates the brand's initial marketing budget, from which a Marketer/ Brand manager can actually develop and execute a strategic plan for the compound.

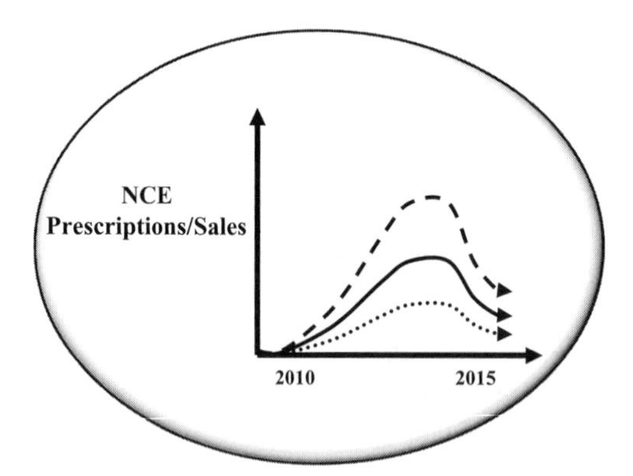

Figure 4 Examples of Stretch, Most Likely, and Worst case scenario outputs from Monte Carlo analysis.

The details of the Brand Plan are specific to each individual situation, but often leverage the institutional knowledge the Marketer/Brand Manager has gained throughout the course of interactions with various functional areas of expertise within the firm.

LIMITATIONS FOUND IN THE EXISTING BRAND PLANNING PROCESS

The creation of a Brand Plan consumes a significant amount of company time and resources—a sound investment if the brand positioning and corresponding strategies are sufficient to meet the stated objectives for the developmental compound. Although the existing Brand Planning process (as described above) is the gold standard approach to developing strategic marketing initiatives and robust evaluations of NCE commercial potential, its effectiveness is hindered in four key ways:

1. **Misapplication of product analogs**—Brand Plans often include an assessment of the launches of analogous products, related to both the details of successful brand positioning and its corresponding financial performance, but *historical analogs are helpful only in the rare case when the underlying causal relationships generating market behavior are similar to that of the NCE in question.* The validity of direct analogies to individual products depends on factors such as product similarity, formulary status, the competitive landscape, patient interaction with the health care system, physician choice criteria, and even epidemiology. Without a means to explicitly capture such dynamics and interrelationships, product analogs are often evaluated based on implicit judgment rather than rigorous analysis.
2. **Failure to leverage the institutional knowledge of cross-functional team members**—While cross-functional coordination has become the standard for the Brand Planning process, the data/information of these team members is often only *implicitly* included in strategic marketing programs or assessment of commercial potential for a compound. Without a common operational framework to incorporate such institutional knowledge, the inputs from various organizational functions are generally not incorporated into the Brand Planning process in a consistent and explainable fashion.
3. **Inherent limitations of static approaches**—Static tools such as spreadsheets are inadequate to capture the complex set of interrelationships present in actual pharmaceutical markets. Spreadsheet models often contain detail about a disease marketplace, but are not good at explaining the dynamic marketplace behavior resulting from the interaction between causal factors—the very relationships vital in developing effective marketing strategies.
4. **Maintaining consistent assumptions when evaluating alternative strategic options**—Brand Plans often make aggregate assumptions about how various marketing strategies might affect the performance of the compound in the marketplace. Typically, the assumptions behind such strategic programs involve an expectation of the effect on the NCE Prescriptions/Sales, as seen in Figure 5.

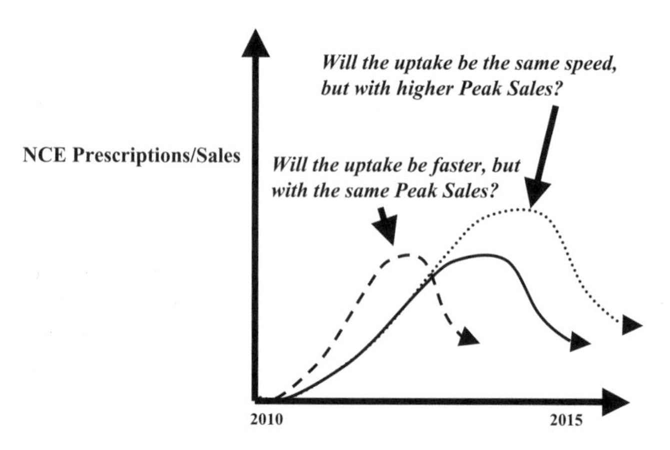

Figure 5 Macro assumptions regarding impact of marketing strategies on compound uptake.

The details driving such macro assumptions, as well as the expected magnitude and timing of their impact on an NCE's expected uptake, are often hidden and unarticulated. Such ideas are often termed as **mental models**— implicit cause-and-effect relationships that are never explicitly defined or sufficiently explained.* Establishing a consistent connection between mental models and their corresponding assumptions is crucial for effective allocation of limited resources, and the development of successful marketing strategies yet is difficult to achieve without an operational framework to make such a link explicit and testable.

LACK OF INTEGRATION BETWEEN FORECASTS AND PRODUCT STRATEGIES

Brand Plan development and its NCE forecast component are designed to leverage knowledge from a cross-functional team, and in a certain sense they do. But because the traditional forecasting process is fundamentally disconnected from the operational realities of the disease marketplace, the Marketer/Brand Manager lacks a tool to quantitatively analyze the effects of the product strategies being developed. *In other words, there is no established link between brand positioning and the corresponding expected sales trajectory of the compound.* Once the NCE is on the market, a Marketer can collect data and calculate a return on investment of various marketing programs. In the prelaunch timeframe, however, rigorously testing the effects of possible strategies is impossible without an operational way to evaluate the expected outcomes of strategic marketing decisions. A Brand Plan based on such untested mental models is often inefficient with respect to its

* Mental models are the basis for many decisions throughout the pharmaceutical industry and are explained more fully in chapter 2.

strategic goals, as it has failed to incorporate the series of causal relationships driving the behavior of the disease marketplace.

The costs of ineffective decision making are significant in the pharmaceutical industry, where the resources required for drug development and marketplace introduction are huge and growing. A recent study by Bain & Company puts the cost of bringing a drug to market at $1.7 billion (6), while the Tufts Center for the Study of Drug Development places the number closer to $802 million (7). These astronomic figures reflect the great risks associated with the development of new compounds and their ability to succeed in the marketplace, and come in three varieties.

Risks Associated with Not Integrating Commercial Assessments with Marketing Strategies

1. **Pursuing an NCE of Limited Potential:** Early commercial assessments often paint a rosy picture that begins to look tarnished if actual market performance of a compound fails to meet expectations. Allocating resources to products whose revenue potential is limited is costly in two ways. First, the direct costs of such efforts are significant and readily apparent. Secondly, each dollar spent on a "sinking ship" product means one less dollar available to invest in other potentially profitable drug development efforts. Forecasting processes such as the traditional one outlined in this chapter are designed to minimize this risk, but new methodologies to help guide and inform this approach can be beneficial as well.
2. **Abandoning an NCE of High Potential:** An understated projection of a compound's commercial value may result in the discontinuation or out-licensing of a product whose revenue stream may turn out to be significant. In the pharmaceutical world, where the revenue from successful drugs must fund new development and cover the costs of failed compounds, failure to "ride out a winner" is a recipe for financial disaster.
3. **Failing to Strategically Position an NCE Effectively:** Without a clear understanding of marketplace dynamics, a Brand Plan may be focusing on low-leverage marketing strategies and consequently fail to maximize a particular compound's potential. Because successful compounds are the exception rather than the rule in pharmaceutical development, maximizing the returns of an NCE as it enters the marketplace is vital to a firm's long-term financial success.

In our consulting experience over the past 10 years, we have seen strategic initiatives and forecasts of commercial potential for a wide variety of diseases, products, and competitive situations. While each of these two components is a key part of the development process for a compound, they remain fundamentally disconnected in the traditional Brand Planning approach. This disconnect continues to hamper effective strategic decision making in the pharmaceutical industry—the costs of which are high and growing. To be truly effective, forecasts should at a minimum consider the operational set of causal relationships driving the behavior of a given disease marketplace, as well as incorporate and estimate the effects of various marketing programs on it. *At the level of a Brand Plan, the expectation of the financial performance of an NCE needs to be operationally tied to the strategies designed to support it.* Without a causal model to link potential marketing actions to expected outcomes, the process of developing, quantifying, and assessing the possible impact of various strategies is done more by gut instinct than explicit and testable analysis. *Stated differently, a commercial assessment of an NCE is not leveraged effectively if its forecast methodology ignores the product strategies associated with it in the Brand Plan.*

Our work with pharmaceutical clients focuses on the explicit integration of the forecasting and strategic planning processes for effective development of Brand Plans. The approach utilizes a Dynamic Modeling framework—the details and value of which will be addressed throughout the rest of this text. Dynamic models do not necessarily replace existing commercial assessment methodologies or supplant traditional Brand Planning processes, but they can establish some operational rigor around the structure and expected behavior of indication marketplaces.

The Dynamic Modeling framework provides a visual language for describing mental models, a platform for explicitly integrating data from various functional areas of an organization, and a simulation tool for understanding and analyzing the causal relationships that make pharmaceutical markets change over time. Over the past decade, we have successfully applied this approach to developing Brand Plans containing product strategies that are fundamentally linked to operational forecasts of pharmaceutical products in a wide variety of situations. The chapters that follow will detail the advantages of the Dynamic Modeling methodology, the ways in which this framework can add value, and insight to the Brand Planning process, and specifics of how such models are built, analyzed, and utilized in the pharmaceutical industry.

SUMMARY

- The challenges facing the pharmaceutical industry in terms of developing, testing, and marketing new drug compounds are daunting.
- A number of inadequate analytic methodologies are used to evaluate pharmaceutical markets and develop strategies therein.
- Strategic decisions are often made based on implicit mental models that are neither testable nor well communicated.
- Input from cross-functional teams to the existing strategic planning process is not efficiently utilized.
- The risks associated with ineffective decision-making processes are huge in the pharmaceutical world.
- Brand Plans should demand an operational way to test the effects of possible marketing strategies on the expected sales trajectory scenarios they are intended to generate.

- The need for an integrated approach of tying commercial assessment of compounds to the marketing strategies designed to support them is met with the Dynamic Modeling approach.

REFERENCES

1. IMS Health Report, 2008.
2. Geographic variation in prescription utilization. Express Scripts, Inc., 2008.
3. FDA Approval Study. BioMedTracker, Sagient Research Systems, 2008.
4. Simon F, Kotler P. Building Global Biobrands: Taking Biotechnology to Market. New York, NY: Simon & Schuster, 2003.
5. Food and Drug Administration (FDA) Orange Book. http://www.fda.gov/cder/ob/default.
6. Landers P. Cost of developing a new drug increases to about $1.7 billion. Wall S J December 8, 2003.
7. Tufts Center for the Study of Drug Development. http://www.csdd.tufts.edu.

A Dynamic Patient Flow Model to Identify Areas to Improve Awareness and Diagnosis in Adults with Attention Deficit/Hyperactivity Disorder

As part of Shire Pharmaceutical's preparation for the launch of their newest ADHD medication, the company recognized that there was significant unmet need among adults suffering from the condition. They hypothesized that many potential patients suspected an attention deficit problem, yet remained undiagnosed and untreated despite regular contact with the health care system. Shire engaged the assistance of a team of system thinkers to simulate the dynamics of pathways to care in the adult ADHD patient population in the U.S. health care system.

The estimated prevalence of ADHD in U.S. adults is 4.4%, yet the diagnosed prevalence is only 1.4%, suggesting a lack of patient awareness and physician under/misdiagnosis. In collaboration with a multidisciplinary group of clinicians and marketers at Shire, AXIA, Lexidyne, and EpiSource constructed a dynamic simulation model of pathways that can lead to diagnosis of ADHD in adults in the United States. This Adult ADHD Patient Flow Model identifies populations of adults with ADHD before diagnosis and diagrams their flows through different stages within and outside the health care system.

This dynamic operational model was constructed in four steps:

1. The results of a literature review were used to identify adult patient populations with ADHD, their pathways to diagnosis, and gaps in clinical management that affect their progress to diagnosis.
2. A pathway-to-diagnosis map was created to provide a visual representation of populations and movements in adults with ADHD prior to diagnosis.

This map was used to refine data requirements and identify assumptions associated with magnitudes of patient pools and movement across those pools on the pathway to diagnosis.

3. A dynamic model was developed and populated with published epidemiologic and clinical data and evidence-based assumptions to fill gaps where data were unavailable.
4. A baseline case was generated to determine the size of patient sources and the flow of those patients through the system, and to evaluate consistency of data and assumptions with known adult ADHD population characteristics.

BASELINE CASE—DISTRIBUTION OF PATIENTS

Patients were mapped to four locations along the pathways to care: outside the medical system (comprising patients either unaware or aware of an issue/suspect ADHD), seeking diagnosis, misdiagnosed, and correctly diagnosed. Differences in flow through the pathways allowed us to identify points of patient accumulations and reductions.

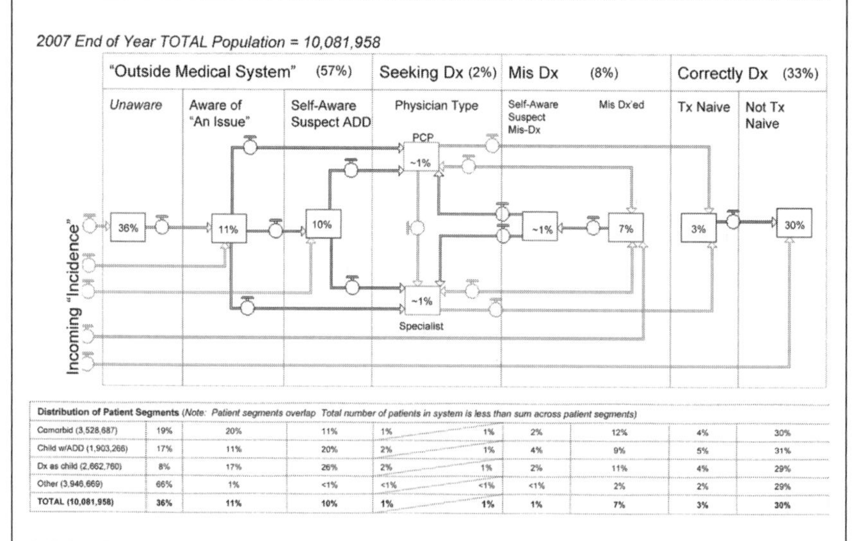

Distribution of Patient Segments (Note: Patient segments overlap Total number of patients in system is less than sum across patient segments)									
Comorbid (3,528,687)	19%	20%	11%	1%	1%	2%	12%	4%	30%
Child w/ADD (1,903,266)	17%	11%	20%	2%	1%	4%	9%	5%	31%
Dx as child (2,662,760)	8%	17%	26%	2%	1%	2%	11%	4%	29%
Other (3,946,669)	66%	1%	<1%	<1%	<1%	<1%	2%	2%	29%
TOTAL (10,081,958)	**36%**	**11%**	**10%**	**1%**	**1%**	**1%**	**7%**	**3%**	**30%**

Adult ADHD pathway-to-diagnosis map: distribution of patients.

- Examining all patients using the dynamic flow model, the baseline case resulted in a 2007 projected end-of-year population of 10,081,958 adults with ADHD in the United States, with 57%, or nearly 6 million adults, undiagnosed and untreated, and therefore, outside the medical system.

- The projected population of adults with ADHD and a mistaken diagnosis was 8% of the total population, or approximately 800,000 adults by year-end 2007. The relatively high percentage of adults with ADHD and a mistaken diagnosis suggests that ADHD in adults may be overlooked in favor of the diagnosis of a different psychiatric comorbidity.
- More than a third of the projected population of adults with ADHD were unaware of their condition, and an additional third were correctly diagnosed, with the majority of these receiving treatment.
- This model identifies specific pathways to diagnosis and points within those pathways that may be actionable, such as "Self-Aware/Suspect ADHD" or "Mistaken Diagnosis." The large numbers of patients in these groups suggest that interventions that enhance patient awareness and promote physician education on the detection of ADHD will have a substantial impact. Such interventions may significantly reduce the number of patients who fail to improve because their ADHD is undetected and untreated.

EXPERIMENTS—IMPACT ON CORRECTLY DIAGNOSED PATIENTS

Using our Adult ADHD Patient Flow Model, we tested the impact of altered flow rates along the pathways to correct diagnosis (simulating results of successful patient/physician educational initiatives) on the number of patients correctly diagnosed over time. We established a baseline case for a 24-month experimental period from January 2008 through December 2009.

- In experiment 1, the Physician Effectiveness Experiment, we examined the impact of decreasing the rate of patient misdiagnosis by 40% and decreasing the length of time to correct diagnosis by 30% over the experimental period.
- In experiment 2, the Patient Activation Experiment, we examined the impact of tripling the rate at which patients outside the medical system seek diagnosis over the experimental period. For both experiments 1 and 2, the primary outcome was the number of patients correctly diagnosed three years beyond the end of the experimental period.
- In experiment 3, we combined experiments 1 and 2 in a parallel experiment.

The results of experiments 1, 2, and 3 indicate that an additional 115,000, 202,000, and 427,000 patients above baseline, respectively, would receive a correct diagnosis of ADHD three years beyond the end of the experimental period. Since ADHD is associated with significant economic consequences [including lost work performance estimated at approximately $4400 per individual (Kessler et al. 2008)], these numbers translate into significant economic savings on a societal level, as well reduced personal suffering for the large numbers of undiagnosed ADHD patients in the United States.

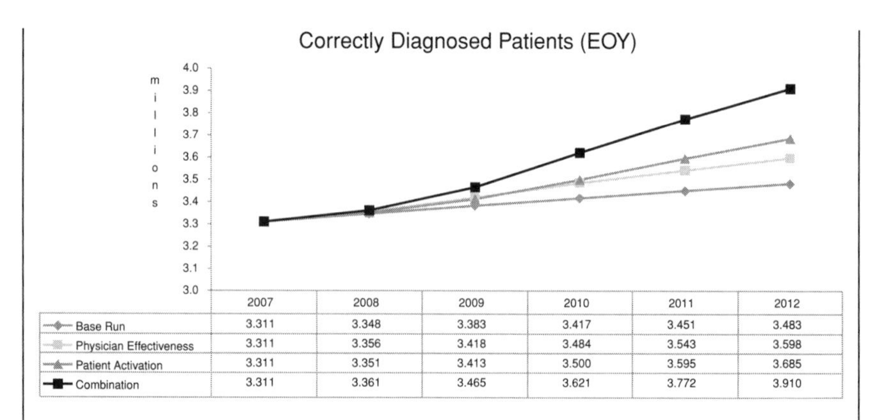

	2007	2008	2009	2010	2011	2012
Base Run	3.311	3.348	3.383	3.417	3.451	3.483
Physician Effectiveness	3.311	3.356	3.418	3.484	3.543	3.598
Patient Activation	3.311	3.351	3.413	3.500	3.595	3.685
Combination	3.311	3.361	3.465	3.621	3.772	3.910

Experiment results: correctly diagnosed patients.

- Our experiments demonstrated that physician-focused interventions ultimately resulted in fewer projected patients with a correct diagnosis than patient-focused interventions because physician-focused interventions cannot drive patients inside the medical system. Conversely, patient-focused interventions resulted in a greater projected number of mistakenly diagnosed patients than physician-focused interventions, emphasizing the need to build diagnostic capacity before accelerating patient movement through the system.
- The results of combining physician-effectiveness and patient-activation interventions demonstrate that the combination can result in greater increases in projected populations of patients with a correct diagnosis of ADHD compared with either intervention alone. This synergy resulted from simulations of higher patient volume entering the offices of physicians and a greater likelihood of the physician making a correct diagnosis.

These findings have been used by Shire Pharmaceuticals to assist in the development of strategies and tactics to aid physicians in improving the diagnosis of adult ADHD. The findings were presented at the 2008 American Psychiatric Association Annual Meeting in Washington, D.C., where questions from physicians in the United States and European countries indicate an interest in development of a screening tool to improve diagnosis of adult ADHD.

REFERENCES

1. Kessler RC, Lane M, Stang PE, et al. The prevalence and workplace costs of adult attention deficit hyperactivity disorder in a large manufacturing firm. Psychol Med. 2008;21:1–11.
2. Willey C, Faraone SV, Peterson S, et al. A dynamic patient flow model to identify areas to improve awareness and diagnosis in adults with attention deficit

hyperactivity disorder. Poster presented at the 2008 American Psychiatric Association Annual Meeting, Washington, D.C., May 3–8, 2008.

3. Faraone SV, Willey C, Peterson S, Stafford J. Diagnosis of Adult Attention Deficit Hyperactivity Disorder: Experimental Results From the Adult ADHD Patient Flow Model. Poster presented at the 2008 American Psychiatric Association Annual Meeting, Washington, D.C., May 3–8, 2008.

2

Dynamic Modeling Approach to Brand Planning: Operationally Understanding Key Marketplace Dynamics

INTRODUCTION

This chapter will cover the following topics:

- Models as useful simplifications of reality
- The structure/behavior link in dynamic models
- Three sectors of Standard Template dynamic model
  - Patient Flow Dynamics
  - Doctor Adoption of pharmaceuticals
  - Treatment Attractiveness of compounds
- Overview of Dynamic Modeling advantages
  - Sanity checks against NCE forecasts
  - Operationalizing compliance and persistence
  - Integrating short-term and long-term forecasts
  - Determining strategic leverage
  - Decision analysis
  - Limiting physics
  - First-to-market dynamics
  - Importance of data

> "All theories are wrong, some theories are useful."
>
> George Box, but often attributed to W. E. Deming.

In our consulting practice, we are sometimes asked if dynamic models are "right" and the answer to that question is a resounding "No." All models are theories, and theories by definition are abstractions from reality. Box and Deming were not being evasive, they were being literal—any type of model, regardless of the methodology or statistical sophistication, is wrong from the beginning.

This concept goes beyond accepted technical measures of model validity such as R-squared or other "goodness-of-fit" metrics—it is a fundamental principle of the real world like gravity, death, and taxes. Models are intended to be simplifications, so beware any model claiming to represent "truth." Wise modeling practitioners leave such matters to the theologians.

Effective models are those that are used to make better decisions, and people use such models everyday—often without realizing it. Did you take an alternate route to work this morning to avoid traffic? Your revised path was the result of a mental simulation that predicted you could get to work faster with a "back roads" strategy. Have you ever reprimanded a child? Doing so is likely the product of a mental model that forecasts the long-term positive effects on the child's future behavior. Brushing your teeth, attending college, exercising all are actions done with the expectation of some derived benefit sometime in the future. In fact, many of the things we do on a daily basis are the results of models—actions or strategies that we expect to have some desired outcome. Although rarely explicitly represented or analyzed, mental models are often the decision-making processes we employ throughout the course of our daily lives.

The situation is not different in the pharmaceutical world, where marketing strategies are developed based on expectations of their effects on drug sales, market penetration, perceived quality, etc. Such strategies are often the product of implicit mental models and are effective when they have their desired impact. But mental models are difficult to communicate, impossible to analyze or quantify, and hard to prioritize. As such, they often result in suboptimal decisions when it comes to strategy development and Brand Planning.

Dynamic models represent an operational way to translate mental models from the implicit to the explicit, allowing for a set of specific business questions to be addressed and analyzed. From a marketing perspective, these strategic questions often center around topics such as the following ones:

1. How can we develop a more effective strategic Brand Plan?
   - Should we grow the market, steal share from competitors, or create new markets for the compound?
   - What key market segments are crucial to the success of the NCE?
2. How can we better leverage the wealth of data and institutional knowledge about the disease indication, patient behavior, and physician preferences?
   - How can our marketing programs impact the observed behavior of these important stakeholders in the disease marketplace?
   - What vital knowledge of marketplace dynamics is missing from our current information sources?
3. How can we sanity check existing commercial assessment methodologies?
   - How likely are various forecasting outcomes?
   - What key assumptions drive the expected sales trajectory of the compound?
4. How can we have an integrated tool to link the results of proposed strategies to impact on the compound commercial assessment?

- Is there a way to evaluate how potential marketing strategies will affect the disease marketplace?
- Can a new methodology take into account the causal effects of various marketing initiatives?

Each of these questions boils down to gaining a better operational understanding of the fundamental processes driving the expected evolution of an indication marketplace—a process that utilizes the **structure/behavior paradigm.**

THE LINK BETWEEN STRUCTURE AND BEHAVIOR

The defining aspect of the Dynamic Modeling approach is its focus on the link between the *structure* of a system and its resulting *behavior*. Dynamic models go beyond statistical correlations or regression algorithms to address the fundamental processes and interrelationships that define system results. Such understanding is essential to effective strategic decision making. To quote Barry Richmond, a visionary in this arena, "By thinking in terms of how a process or system really works (i.e., its 'physics'), we have a much better chance of understanding how to make it work better!" (1).

The Dynamic Modeling framework forces an understanding of the structure of an indication marketplace in order to formulate a better, more effective set of strategies for operating within it. Knowing the uptake trajectories of previously released compounds in the indication or isolated data on epidemiology, treatment paradigms, physician attitudes, and competitive products is not enough. Rather, an integrated approach that combines the effects of these operational factors can not only explain the sales trajectories for product analogs, but also provide a means to quantify such key strategic questions, such as the following ones:

- How many patients are newly diagnosed each year?
- How many patients are switching treatments?
- What drugs are they switching from and to?
- How are changing patient demographics and/or epidemiology likely to affect the key patient dynamics?
- What stages do physicians progress through in their acceptance of a new compound?
- How do marketing levers affect the speed at which that diffusion takes place?
- What are the currently available treatment options in the indication?
- How do these therapies compare to one another and the NCE?

Answering questions such as these depends on a clear understanding of marketplace dynamics and the structure/behavior paradigm, resulting in effective and successful Brand Plans and corresponding marketing strategies. Structure and behavior links are pervasive in the pharmaceutical industry, and the Dynamic Modeling approach can serve to frame and analyze the sets of interrelationships that ultimately drive marketplace evolution.

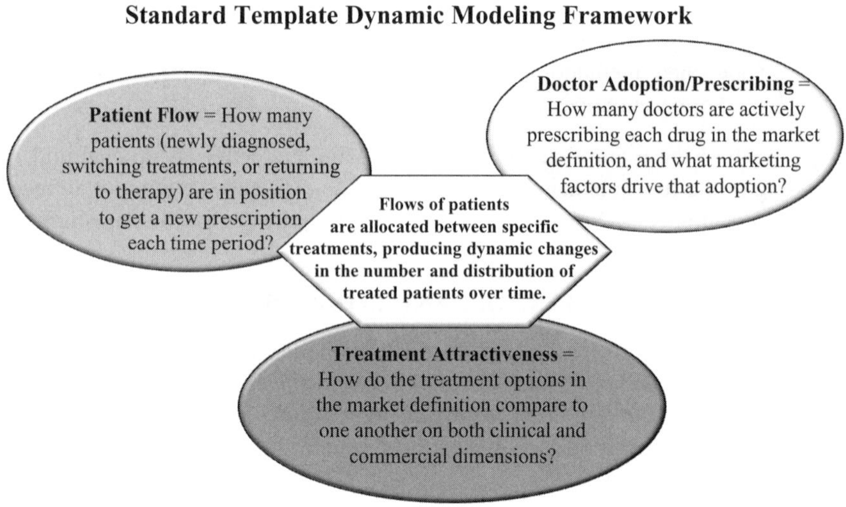

Standard Template Dynamic Modeling Framework

Patient Flow = How many patients (newly diagnosed, switching treatments, or returning to therapy) are in position to get a new prescription each time period?

Doctor Adoption/Prescribing = How many doctors are actively prescribing each drug in the market definition, and what marketing factors drive that adoption?

Flows of patients are allocated between specific treatments, producing dynamic changes in the number and distribution of treated patients over time.

Treatment Attractiveness = How do the treatment options in the market definition compare to one another on both clinical and commercial dimensions?

Figure 1 The three sectors of the Standard Template Dynamic Modeling framework.

THREE KEY SECTORS OF THE DYNAMIC MODELING OF PHARMACEUTICAL MARKETPLACES

We have developed what we call the Standard Template dynamic model as shown in Figure 1 that answers some common strategic questions by combining aspects of Patient Flow dynamics, Doctor Adoption of pharmaceutical products, and the perceived Treatment Attractiveness of the treatment options within a disease marketplace. These components are populated with data from epidemiology and a variety of physician and patient databases to ensure a robust representation of actual market dynamics. Merging these three structural pieces into a dynamic model determines what drugs patients receive, how long patients remain on treatment, how patients transition between therapies, how rapidly doctors accept new forms of treatment, and the evaluation of available treatments in the marketplace. In the Standard Template, these three key components come together to form an integrated model that captures and quantifies the important dynamics that operationally determine how pharmaceutical markets behave over time. Leveraging this knowledge in an operational simulation environment provides the means to test the effects of various strategic options.

Patient Flow

Product-based businesses rely on giving customers the opportunity to use a particular offering, and a number of different frameworks exist for analysis of consumer goods markets (2). The operating principle for pharmaceutical products is getting patients to try and continue to use an individual drug therapy. Pharmaceutical firms often collect or purchase data on the number of prescriptions written and/or filled,

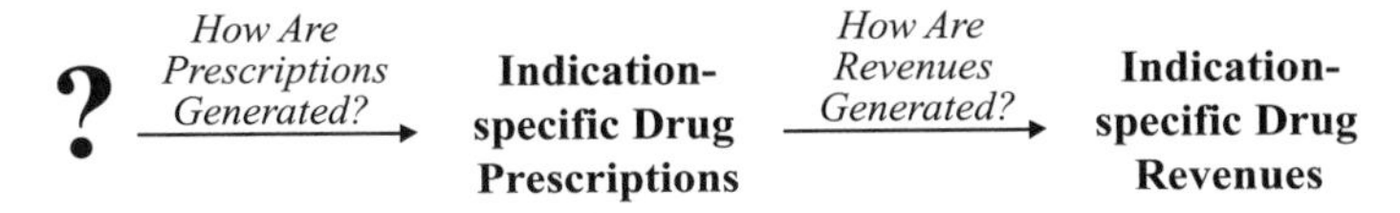

Figure 2 Traditional way to look at prescription/revenue generation in pharmaceutical marketplaces.

as prescriptions can be translated into a revenue estimate for a particular drug in a given indication. Yet rarely does the effort ask the fundamental question of the driving forces behind prescription generation, as shown in Figure 2.

As was discussed in chapter 1, early commercial assessments of NCEs often employ statistical analysis to produce projections of the numbers of prescriptions in an indication marketplace through time. But from a causal standpoint, forecast methodologies do not address the underlying factors that determine how prescriptions are created in the marketplace. Operationally, prescriptions are generated by patients as they

- initiate prescription treatment for the first time,
- switch from one treatment to another,
- return to treatment after having been not treated for a certain time period, and
- refill their existing prescriptions periodically.

These treatment opportunities represent the flow of patients in a given indication over a specific period of time. There is a famous saying that the education at MIT is like trying to take a sip from a fire hose, which in the end is better than taking a large gulp from a smaller information stream at a less prestigious university. The same principle applies in a pharmaceutical context, but relates to the flow of patients moving within a marketplace and the number of those patients captured by individual compounds. *Operationally, the volume of patient flows and their allocation between various treatment options in a marketplace fundamentally determine how the magnitude and associated shares of patients/prescriptions/revenues will change over time.* From a strategic standpoint, the questions become, "How big are the relevant patient flows, how many patients might a particular NCE capture, and what strategies would help capture them?" The Standard Template model defines the magnitude of these treatment dynamics under the heading of Patient Flow, and establishing such an operational structure for prescription generation and the resulting analysis is extremely important for effective strategic planning.

Case Study: Operationalizing Expected Sales Trajectory

A company is launching an NCE into a disease marketplace of 1,000,000 currently treated patients, and wants to know how many patients might be captured during a critical three-year launch period. Analogs and an early commercial assessment have indicated that a 20% patient share is possible at three years postlaunch—somehow the firm must capture (and keep!) 200,000 patients over the next three

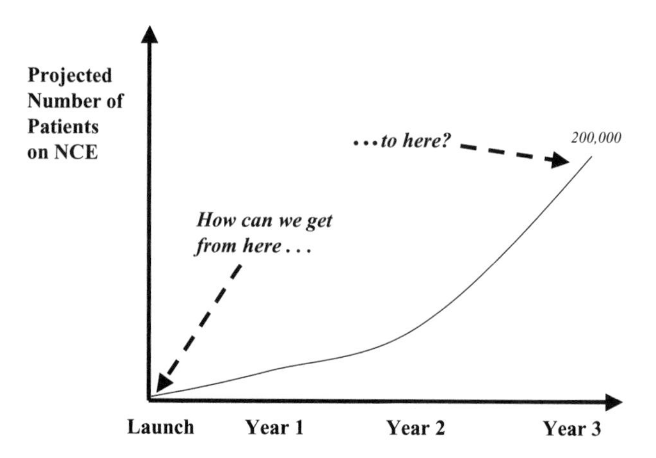

Figure 3 Operational understanding of patient projections.

years. Figure 3 depicts the key operational question behind such an expected sales trajectory.

Traditional projection methodologies typically say nothing about the patient dynamics that ultimately drive the type of uptake curve shown in Figure 3. Operationally, this accumulation of patients results from capturing and keeping a combination of patients who are switching treatment, reinitiating treatment, and others who are completely new to treatment. Most forecasting efforts imply patient numbers at given time points, such as 200,000 patients at the three-year mark, but are silent as to the operational assumptions behind such projections. The more important question from a strategic standpoint is whether such an expected trajectory is reasonable given observed marketplace dynamics surrounding the flow of available patient opportunities.

Suppose a series of market research projects indicated that only 10,000 patients are new to treatment every year, another 20,000 switch therapies, and only 5000 patients return to treatment over a year's time. As Table 1 indicates, even assuming the new compound captures every single one of those patients and

Table 1 Operationalized Mathematics Behind Example Marketplace Dynamics

Patient flow	Total patients/ year	New drug captures	For how many years?	Total patients on NCE
New to treatment	10,000	100%	3	30,000
Switching treatment	20,000	100%	3	60,000
Reinitiating treatment	5,000	100%	3	15,000
Total	35,000			105,000

keeps them indefinitely, the original estimate of 200,000 patients at three years postlaunch begins to look incredibly optimistic.

Note that from an operational standpoint, Table 1 suggests that the expectation of 200,000 patients is simply not achievable. But what about the effect of strategic marketing campaigns? Could not effective programs increase the volume of patient flows in the marketplace in order to meet the projected patient numbers? Marketing strategies are intended to do just that, but even back-of-the-envelope numbers such as those shown in Table 1 begin to provide some boundaries for such assumptions. Note that marketing campaigns would have to double the respective patient flows for the entire three-year period to even have the possibility of hitting the 200,000 patient mark. This type of analysis, based on quantifiable and operational dynamics regarding patient opportunities, allows the reasonability of assumptions to be examined. As importantly, these types of questions are even rarely asked without the operational perspective of the Dynamic Modeling process.

Dynamic models operationally define the flows of patients who represent opportunities in a given marketplace. Starting in chapter 4, we will delve more specifically into understanding and modeling these key patient dynamics. In general, the Patient Flow component recreates the structure and magnitude of patient movements within a simulation environment representing an indication marketplace. This type of operational examination of market dynamics becomes a useful sanity check for traditional forecasting efforts. In addition, it can help inform and test key marketing strategies being considered for an NCE's Brand Plan.

The differential behavior of various patient segments can also be included in the Patient Flow sector as well. For example, patients may be further disaggregated by severity of their disease: mild, moderate, and severe. A drug may only be useful for mild or moderate patients, so understanding patient flow behavior by severity is crucial in making sound marketing decisions regarding product positioning and strategy. Other types of segmentation, such as by age or gender, may be important as well. Epidemiologic concepts such as incidence, prevalence, diagnosis, and treatment rates often vary across these dimensions. The Patient Flow structure is fully flexible in terms of segmentation, allowing for a more accurate representation of differential dynamics of various patient groups.

Doctor Adoption

In many pharmaceutical drug markets, historical data can be analyzed to determine how quickly drugs in a particular indication were adopted by prescribing physicians. Prospectively, market researchers often collect data on awareness of drugs that have yet to be launched and conduct surveys on doctors' expectations regarding their future prescribing patterns. This information, while useful, does not explicitly capture the mechanisms at play in both past and future adoption of pharmacologic treatments by doctors, nor does it indicate how such dynamics interact with other parts of the marketplace to affect overall market performance.

As we will examine in chapter 8, the Dynamic Modeling framework can be applied to issues surrounding Doctor Adoption in order to accomplish the following:

- Understand the operational physics driving adoption of pharmaceutical products.
- Determine the importance of certain marketing levers in driving that adoption.
- Relate the aspects of physician adoption to components of patient dynamics and various treatments in a competitive landscape.
- Develop strategic forecasts based on an integrated view of marketplace dynamics.

Dynamic Modeling utilizes a physician adoption structure that is supported by long-standing statistical approaches to product diffusion and can replicate the historical behavior of various drug analogs for a particular indication. This analysis establishes a visual, operational representation of the doctor adoption process, and creates a powerful simulation tool to test the effectiveness of various physician-marketing strategies. As chapter 8 will show, published marketing literature can provide estimates regarding the effectiveness of various marketing expenditures specific to the pharmaceutical industry. These diffusion factors are easily translated into the Doctor Adoption sector of the Standard Template framework, quantifying the effect of various physician-marketing strategies on the overall marketplace both now and in the future.

Doctor Adoption frameworks generally begin with aggregated analyses of physician behavior that can then be disaggregated into more detailed doctor segments along lines of specialty, adoption propensity, and even disease-specific characteristics. For example, general practitioners may be accounted for separately from specialists in an indication, including word-of-mouth or referral patterns between these two groups of potential treating physicians. Data analysis or attitudinal studies may reveal that certain doctors tend to adopt new products faster than their peers, suggesting a further model disaggregation according to adoption propensity. Dynamic models often include the adoption of drugs specific to certain patient segments—a feature that is especially useful for indications in which, for example, a new treatment may be readily accepted by physicians for their most severe patients, but not for patients suffering from only mild symptoms. The ability to incorporate and test the adoption behavior of different physician segments into a dynamic model provides a more realistic and detailed picture of how effective marketing strategies related to physician acceptance of products can best be developed.

Treatment Attractiveness

Consumers on a daily basis make purchasing decisions based on perceptions of how products stack up against one another. Many factors may affect perceived product attractiveness: price, function, look, style, status, etc. And while products vary across indication and consumers are not identical in their choice criteria, understanding the process through which purchasing decisions are made is vital when analyzing consumer markets.

A similar evaluation process takes place when physicians make prescribing decisions for patients with a particular disease. A set of product criteria is used to determine which drug is most appropriate to prescribe in a given situation. Dynamic Modeling allows the concept of treatment attractiveness to be applied to therapy options in a particular indication, allowing the Brand Plan team to determine the relative utility of treatment options in an indication. More importantly, the Dynamic Modeling process can help determine the degree to which these differences affect prescribing decisions in the marketplace.

As we will see in chapter 10, Treatment Attractiveness is an overall metric of utility that can be a function of product attributes such as safety, side effects, price, etc. The evaluation of attractiveness can also be tied to specific patient segments or physician specialties in order to more closely replicate differential evaluations in a complex marketplace, providing an additional level of analysis and insight within the Standard Template.

OVERVIEW OF DYNAMIC MODELING ADVANTAGES

Standard Template dynamic models represent a departure from traditional approaches, but have a number of distinct advantages that make their application a vital part of effective Brand Planning.

Sanity Checks Against Other Forecasts

Many pharmaceutical companies generate an early commercial assessment of an NCE for planning and strategy development. As chapter 1 demonstrated, for example, traditional forecasts may estimate the number of patients, volume of prescriptions, or the resulting level of sales for a particular new compound in each year after its launch. However, these projections may not be realistic in the current market environment and Dynamic Modeling provides a useful comparison in order to check the feasibility of such forecasts.

Case Study: Forecast Reasonability

One of our clients developed an NCE forecast indicating a 10% patient share three years postlaunch, which some members of the Brand Plan team felt was overly optimistic. A gut check told team members that this evaluation of commercial potential was not achievable, and the resulting Brand Plan was setting itself up for failure. A simple dynamic model was developed for the marketplace as a comparison to the existing forecast, and the team discovered that what management believed to be a relatively innocuous goal of 10% patient share turned out to be incredibly optimistic given the dynamics of the marketplace. This process allowed the team to provide upper management with a sanity check of expectations and adapt a forecast that was both more realistic and achievable.

Such sanity checks can be extended to the arena of treatment attractiveness to determine how good an NCE must be to attain a desired market or patient share. The Standard Template takes inputs regarding the comparative utility of treatment options and calculates resulting patient shares and associated product revenues over time. The model can also be used in reverse to determine how good a new compound must be in order to attain a certain market share target. These types of analyses often prove useful in the early stages of drug development when "go, no-go" decisions are being made.

The epidemiology factors of a disease are another area where dynamic models provide important insights and clarity for marketing teams. The explicit structure of disease progression and the associated epidemiology metrics are rarely included in traditional forecasts. However, the dynamic model framework clearly accounts for these factors by showing the relative magnitude of patient dynamics as they apply to the epidemiology of a disease (see chap. 4). Metrics such as true incidence, diagnosed incidence, point- and lifetime prevalence, and recovery/relapse rates are all typical outputs of the Standard Template. Identifying the magnitudes of and relationships between these epidemiology concepts provides a useful perspective on a disease and its associated market.

Case Study: Misguided Strategy Based On Inaccurate Understanding of Epidemiology

For example, one of our clients was formulating a strategy of market penetration based on capturing newly diagnosed patients. Their spreadsheet model was a "bottom-up" approach based on projected number of doctor visits, the number of visits at which the medical condition could be identified, likelihood of a prescription, etc. The detailed analysis indicated a strategic opportunity to tap into large flows of diagnosed incidence, and the resulting forecasts showed a very quick uptake of the NCE based on this strategy. However, the static approach did not account for the relationship between incidence and prevalence captured in an integrated epidemiology framework. A very basic dynamic model, populated with the client's derived epidemiology data, showed a huge inconsistency in incidence/prevalence assumptions. The incidence rates calculated by the bottom-up approach implied that the prevalence of the disease would nearly double in only three years. This quick sanity check using Dynamic Modeling principles allowed the client team to truly understand the epidemiology of the disease and abandon a doomed strategy of pursuing newly diagnosed patients—a strategy based on a flawed understanding of key epidemiology concepts.

Compliance/Persistency

Compliance and persistency are typically combined in traditional analysis of pharmaceutical markets into one metric to account for patients who do not take all of their prescribed medications. We have found, however, an important psychologic and dynamic difference between these two concepts. In the Patient Flow sector of the Standard Template, compliance is a continuous metric to account for patients who continue to fill their prescriptions but do not take 100% of their prescribed dosage. Noncompliant patients are a continuing source of revenue, although at a reduced rate due to the corresponding extension of time between prescription refills. In Dynamic Modeling parlance, persistency relates to how long patients stay on any form of treatment before discontinuing prescription medication entirely. Non-Persistent patients are no longer revenue –generating, as they have ceased to be treated by any prescription therapy.

Separating these two concepts allows the Patient Flow framework to operationally address some important questions regarding patient dynamics and resulting effects:

- How do assumptions regarding compliance affect overall revenue projections?
- What are the market-level persistency metrics?
  - How long do patients stay on treatment?
  - Is persistency different for different treatment options?
  - How many patients are Non-Persistent at a point in time?
- What do persistency metrics in the indication suggest regarding strategic planning?

As chapter 4 will demonstrate, accounting for compliance and persistence separately in the Standard Template allows for a better understanding of the patient dynamics around these key issues, resulting in better strategic marketing plans to address them.

Integration of Short- and Long-Term Forecasts

Many pharmaceutical firms have short-term and long-term forecasts created by separate teams using different and even conflicting methodologies and assumptions. Reconciling these approaches is often impossible, resulting in confusion and inconsistency at various levels of the organization. In contrast, the Dynamic Modeling framework is not dependent on a specific time horizon, and as a simulation-based approach is able to calculate outputs in any useful time frame. A dynamic model can be programmed to show monthly dynamics that are especially important at the very early stages of a drug launch. These monthly metrics can be aggregated into quarterly or yearly outputs, which are more appropriate for long-term forecasting and/or strategy development. Having one tool using a single Dynamic Modeling approach and populated with identical data ensures internal consistency between short-term and long-term forecasting efforts.

Case Study: Conflicting Forecast Methodologies

In the first few months of launch of a new compound, one of our client teams was having difficulty reconciling their short-term forecasting with the previously established commercial evaluation of the drug's long-term potential. The methodology to address monthly prescription forecasts was a complex algorithm of physician acceptance based on marketing expenditures, correlated to historical data on launches of similar compounds. The long-run revenue projection, however, was based on aggregate trends in market size and estimations of penetration curves over a three- to five-year interval. The client knew that the long-term performance of the drug would simply be a result of a series of short-term performance metrics—just as the flight path of trans-Atlantic airplane is a function not only of the plane's takeoff but also of the subsequent changes in altitude, speed, and direction. But due to incompatible projection methodologies, the team had no way to analyze the long-run implications of the monthly prescription forecasts they were developing.

A dynamic model was created to address this issue, with a consistent methodology for both monthly performance metrics and their resulting long-run implications in terms of market share. The resulting simulation allowed the team to reconcile their existing two approaches in a single framework, providing them with a single tool to analyze the results of various input assumptions on different time horizons.

Chapter 7 will address the issue of short-term forecasting in more detail, and various chapters throughout this book will investigate various Dynamic Modeling frameworks for converting patients into revenues/prescriptions.

Determining Strategic Leverage

Operational representations and quantifications of marketplace dynamics provide a means to rigorously test potential strategies to find points of leverage. A dynamic model can establish a set of base case outputs, based on a defined structure and associated data inputs, which can then be tested to determine their strength of influence. Brand Plan teams are often eager to understand the relative impact of these variables in order to formulate effective marketing strategies.

For example, a marketing team could ask the dynamic model simulation to individually change each input variable one at a time by some factor—typically 10% above or below the variable's base case value. Suppose the rate at which patients switched from an existing treatment option was 4% per month. In a simple sensitivity analysis, the dynamic model would test that variable at both 4.4% and 3.6% and record the resulting behavior of the system. By testing each input variable individually, the Brand Plan team can generate a list of possible leverage points that appear to have the greatest impact on the indication marketplace being modeled.

Case Study: Sensitivity Analysis to Determine Market Leverage

One of our clients developed a Standard Template dynamic model to replicate the marketplace for a particular psychiatric disorder. A sensitivity analysis was then performed to investigate areas of potential strategic leverage within the indication. Figure 4 is an example of how basic sensitivity results are typically displayed, showing the incremental number of patients on an NCE at a particular future time point versus the Base Case.

Sensitivity analysis can be performed in a number of ways; the example in Figure 4 represents a basic sensitivity in which each dynamic model input variable is changed by some constant percentage in isolation over the course of

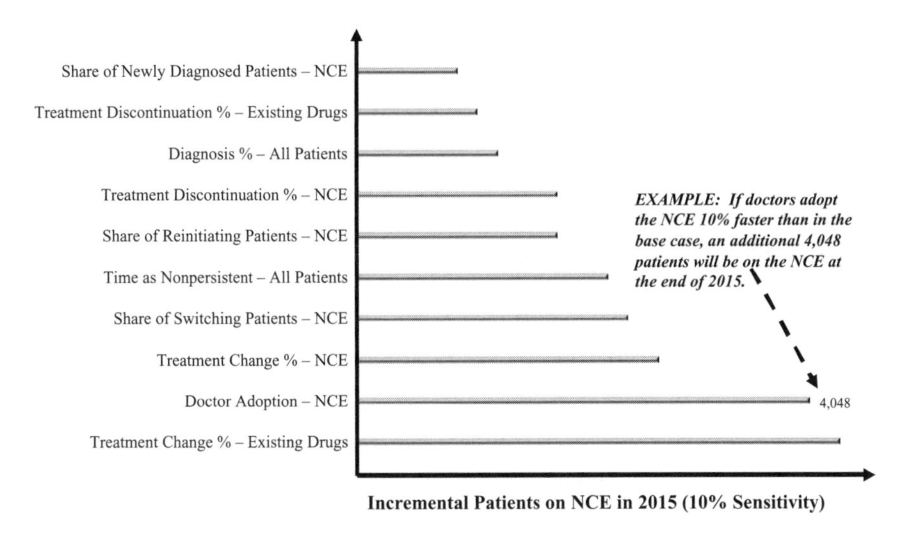

Figure 4 Example for basic sensitivity analysis.

the simulation. The flexibility of the Standard Template allows for a wide range of dynamic sensitivity analyses to be performed and analyzed:

- *Basic—changing the magnitude of each model input by some constant.*
- *Magnitude—testing the differential effect of changing an input variable by differing amounts.*
- *Timing—varying the time at which a change in an input variable takes place.*
- *Combinations—checking combinations of variables and strategies.*
- *Optimization—running multiple simulations and testing all combinations of strategies to find the optimal solution.*

With an integrated dynamic model, sensitivity analyses shows how the entire indication marketplace will be affected over time—something not easily or commonly done in other analytic approaches. Sensitivity analysis and optimization represent a useful way to quantify the possible effects of strategies created during the Brand Plan process.

After the sensitivity analyses were complete, our client wanted to identify the cost/benefit tradeoffs among various marketing strategies. We helped them develop a Strategic Sensitivity Grid, as shown in Figure 5.

The upper right quadrant of the matrix shown in Figure 5 represents high-impact, low-cost options that provide significant leverage in the marketplace. Plotting the results from various sensitivity analyses on such a grid helped the client's marketing team analyze and prioritize their Brand Plan to effectively leverage the observed dynamics in the indication marketplace.

Chapter 7 will provide an extended real world example of how sensitivity analysis and the associated Strategic Sensitivity Grid can be used for strategy

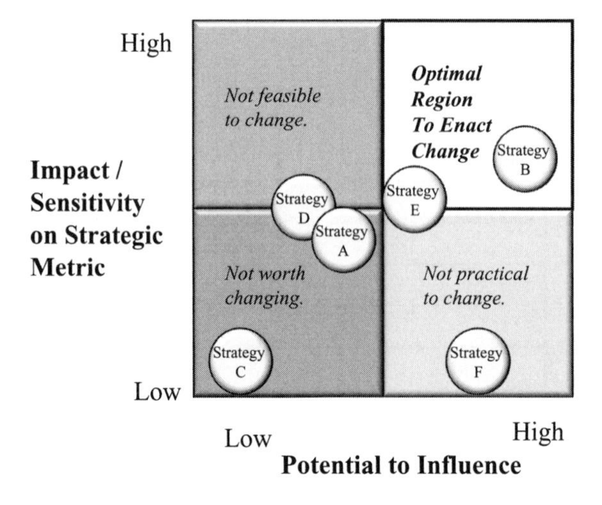

Figure 5 Example for Strategic Sensitivity Grid.

development, and chapters 12 and 14 will detail how sensitivity analysis can be used for strategy development.

Decision Analysis

Marketing teams commonly employ a Strategy Table comparing various strategic options designed to reach their operating objectives or target forecasts. This list of possible strategies is usually organized by potential marketing initiatives and their expected financial return for the compound. Brand Plan teams often spend days debating the pros and cons of various strategies, the reasonableness of their associated financial metrics, and the pitfalls in their tactical implementation. The results of such evaluations are a small group of strategies in the form of a Brand Plan designed to meet the financial goals of the compound in the short and long term.

The drawback of decision analysis based on a common Strategy Table is that *the assumptions behind various strategies that ultimately drive associated financial performance of the NCE are rarely explicit and hence not subject to scrutiny or debate*, as shown in Figure 6.

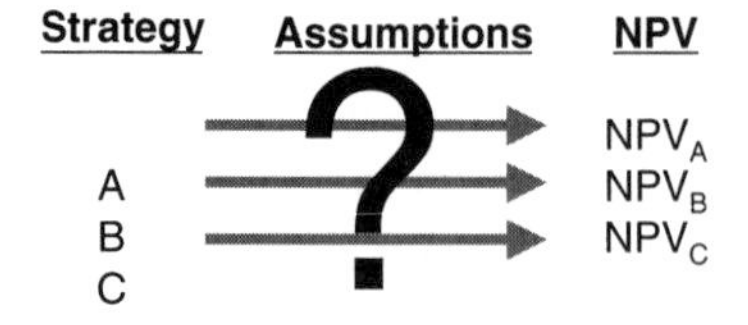

Figure 6 Fundamental disconnect in the typical decision analysis process.

Decision analysis sessions can sometimes devolve into turf wars, in which political clout and strength of personality substitute for rigorous analysis and open discussion/evaluation of strategic options. In our experience, marketing teams frequently agree to the strategies behind a Brand Plan without fully understanding the true nature of the assumptions behind it. In the parlance of Dynamic Modeling, these situations arise from a failure to understand the link between structure (the components, relationships, and assumptions driving a particular strategy) and behavior (the expected result of that strategy).

The Dynamic Modeling methodology offers an intuitive and visual framework for establishing an operational link between possible strategic actions and their expected outcomes. Utilizing such a dynamic model allows marketing teams to undergo the process of decision analysis in a way that *explicitly* captures the knowledge of various team members and company functions. The resulting simulation tool provides the means to rigorously compare the results of various strategic options. Doing so elevates the process of evaluating a set of Strategy Table possibilities from argument to assessment, as team members are able to debate not just a set of strategic expectations but also the detailed assumptions driving each. In this manner, the Dynamic Modeling process leverages team knowledge and information more effectively, resulting in better strategic decision making in the world of Brand Planning.

Limiting Physics

As was discussed in chapter 1, traditional early commercial assessments of NCEs often fail to account for the dynamic complexity that ultimately drives behavior in an indication market. The Dynamic Modeling process, however, captures a series of operational dynamics driving marketplace evolution. This analysis often shows behavioral limits imposed by the dynamic structure—what we like to call the "physics" of the system. Limiting physics dictate practical boundaries for how an indication marketplace can be expected to evolve, and forecasts that exceed these constraints or the marketing plans built from them are in direct conflict with the underlying dynamics at play. Failure to account for or recognize such limiting physics can result in flawed or even fatal strategies.

Case Study: Impact of Limiting Physics on Strategy Development

One of our recent clients had a brand plan marketing strategy concentrated on pulling patients directly from a competitor. There were a number of existing drugs in this marketplace, but our client's new compound and that of a major competitor (Brand X) represented major advancements in treatment of the disease. The client's launch strategy was to gain market share by capturing patients directly from Brand X.

A dynamic model was built to address the feasibility of this strategy by analyzing the current behavior of Brand X's patients. The data analysis and resulting simulation model outputs demonstrated that patients were rarely discontinuing use of Brand X. Even if marketing efforts significantly increased Brand X's switching rate and/or allowed the client's NCE to capture all of those switching patients, the treatment opportunities around Brand X's patient flow dynamics were insufficient to meet the client's strategic goals. Ultimately, the team used the dynamic model to formulate a strategy based on capturing the large flow of formerly treated patients as they returned to treatment. By identifying the limiting physics of the marketplace, the Dynamic Modeling process allowed the team to refocus their marketing efforts to areas that would meet strategic expectations.

Dynamic Modeling can help marketing teams understand the operational physics driving marketplace behavior and which strategies can take best advantage of those dynamics. By focusing effort on areas of high leverage as determined from a Dynamic Modeling exercise, Brand Planners can better use their resources and pursue more effective strategic opportunities. A number of examples throughout the remainder of this text will demonstrate the importance of operationally understanding the limiting physics in pharmaceutical marketplaces, and how strategic plans should be developed accordingly.

First-to-Market and "Me-Too" Drugs

One of the most powerful aspects of the Dynamic Modeling methodology is the incorporation of time and the associated delays involved in launching a new compound. Because dynamic models are simulation based, they have the flexibility to "step forward" in time according to any specified interval. Explicitly capturing the time dimension of key market dynamics allows marketing teams to test the implications of getting to market first versus coming in later as a me-too drug, allowing the importance of entry position into the marketplace can subsequently be evaluated and analyzed.

Case Study: Evaluating Market Entry Order

One of our clients had two developmental compounds for a particular disease indication. The first NCE appeared to be much less efficacious than existing treatment options, although a low side-effect profile was potentially appealing to patients with tolerability issues. The second compound was scheduled to launch three years later and was basically a me-too drug; its profile indicated similar efficacy, side effects, and safety metrics as other therapies currently on the market.

Our client needed to decide their strategic direction in this particular marketplace vis-à-vis their possible offerings and entry timings. Which of the two compounds should be brought to market, or would launching both drugs be a more effective strategy? In effect, there were five separate options to consider regarding market entry of the two NCEs, as shown in Figure 7.

A Standard Template dynamic model was developed to allow the team to test various entry strategies and examine how the indication marketplace might

A ⟶ Alone, abandon/outlicense B
A ⟶ B
B ⟶ Alone, abandon/outlicense A
B ⟶ A
A + B simultaneously

Figure 7 Options of market entry order for two NCEs.

respond accordingly.* Running all five scenarios through the resulting simulation allowed the team to systematically compare their strategic options and evaluate the possibility of having multiple drugs on the market. Because the Dynamic Modeling methodology established a consistent framework to capture both the timing of and interaction effects between various market entry scenarios, the results provided the insight necessary to make an informed strategic decision regarding the future of the compounds.

Chapter 13 will investigate market entry dynamics in greater detail, including quantification of network effects, which might translate into significant first-mover advantage.

The Role of Data in Dynamic Modeling

Although pharmaceutical companies spend millions of dollars on collecting and storing data of various types, *the real competitive advantage lies in properly using and analyzing this information effectively*. As Brand Plan teams develop a dynamic model structure, the resulting picture of the marketplace helps focus efforts on the most relevant and useful pieces of information necessary both to complete the model and to answer some important strategic questions. In Dynamic Modeling projects, data collection becomes an integral part of the overall decision-making process that exists in concert with, rather than isolation from, overall strategic objectives.

Brand Planners usually have a wealth of information collected from various functional departments, but this knowledge often lacks an operational platform to integrate and utilize it effectively. A dynamic model can leverage both the information that exists within the organization as well as the talents of various functional areas to gather more information. On occasions, this process has even identified multiple functional areas that have been collecting (and paying for!) exactly the same information—clearly not the best use of scarce time and resources.

The Dynamic Modeling process helps prioritize a number of pertinent and actionable questions regarding data collection:

1. *What important information is needed to complete the model?* The Dynamic Modeling process focuses data-collection efforts on areas that will ultimately inform the strategic direction for a compound. This structured approach helps

* Abandonment versus outlicensing issues can also be evaluated using the Dynamic Modeling approach. (see chap. 16).

identify places where good information is not necessary, and perhaps more importantly, areas in which current and valid data are vital to effective decision making.

2. *Are there significant gaps in the team's knowledge base?* Dynamic models often identify significant deficiencies in the marketing team's knowledge base regarding a disease and/or the market that has evolved around it. The process may pinpoint aspects of the disease market requiring further data collection to enhance the team's knowledge and create competitive advantage for the compound. Without such a structured procedure for evaluating data needs, these knowledge gaps often go unidentified.
3. *How should relevant data be collected?* The Dynamic Modeling process provides a rigorous format to data-collection efforts that ensures internal consistency and external validity. Chapter 6 will address the specifics of such data formats, including epidemiology, treatment algorithms, and longitudinal patient-level data analysis. These data-collection formats often provide insight into marketplace dynamics and the marketing strategies designed to leverage them.
4. *How should ongoing data collection be structured?* Initial data-collection efforts rarely locate all the necessary information to populate a dynamic model, but the process often identifies specific areas in which further research would provide a strategic marketing advantage. We often class ongoing data collection needs into three categories:
   - Existing data sources/formats that should be updated periodically and incorporated into the dynamic model.
   - Data that have not yet been identified/collected but that simulation results indicate have a huge impact on market dynamics.
   - Tracking metrics to determine if the strategies of the Brand Plan have had their desired effect in the marketplace.

Data represents the engine of a dynamic model, allowing defined structure to be simulated to produce expected behavior. As examples throughout the rest of the text will show, the Dynamic Modeling methodology has some precise ways to estimate, collect, and utilize data in order to develop more effective Brand Plans within the pharmaceutical industry.

SUMMARY

- All models are wrong by definition, but dynamic models have the advantage of incorporating operational physics in explaining marketplace behavior.
- The Standard Template incorporates aspects of Patient Flow dynamics, Doctor Adoption of newly released pharmaceuticals, and Treatment Attractiveness metrics of indication-specific therapies.
- Dynamic models offer several advantages to strategic planning that focus on an operational understanding of marketplace dynamics and a scenario tool

creating cause-and-effect linkages between potential marketing initiatives and their expected outcomes.

- Later chapters will detail how the Dynamic Modeling approach can be used in the areas of compliance/persistence, determining points of strategic leverage, evaluating the effects of order of market entry, and understanding the role of data in effective decision making.

REFERENCES

1. Richmond B, Peterson S. An Introduction to Systems Thinking. From the Users Manuals for the iThink Software. Lebanon, NH: Available from iseesystems, Inc., 1996.
2. Lilien G, Kotler P, Moorthy KS. Marketing Models. Englewood Cliffs, NJ: Prentice-Hill, 1992.

3

The Dynamic Modeling Language

INTRODUCTION

This chapter will cover the following topics:

- Dynamic Modeling 101
  - Example 1: Account Balances
- Dynamic Modeling Calculations
  - Example 2: College Student Population
- Advantages of Dynamic Modeling
  - Example 3: Population Demographics

Simple ideas can often have a huge impact in the business world, particularly when established analytic approaches are made accessible through intuitive new frameworks. The Total Quality Movement started by Deming and advanced by others took existing statistical methods and made them business friendly through easy-to-use tools. Michael Porter took principles from industrial organization and made them accessible using an intuitive, actionable framework. Even Wall Street's focus on shareholder value stems from the work of Rappaport and others to bring economic principles to the world of finance in a usable way.

In a similar fashion, the principles behind the Dynamic Modeling approach have their home in differential equations and control theory, and have been used for decades in the world of engineering. But it was not until these principles were made accessible through an elegantly simple framework, an associated set of simulation tools, and an expanding availability of data that the methodology has become an efficient tool in the business world. The Dynamic Modeling approach is driven by and leveraged from well-established analytic principles, but couches them in an easily digestible framework that resonates even with technophobes. The methodology is not difficult to conceptualize, complicated to explain, or hard to implement—in fact, many of our clients have somewhat of an epiphany when first exposed to this framework. A frequent comment is "I never thought about this issue in that way before, but it makes perfect sense." That type of clarity and insight, coupled with the degree of analysis which the Dynamic Modeling methodology provides, has been the cornerstone of our consulting practice over the last decade, and continues to drive our work with clients in the pharmaceutical industry and beyond.

The Dynamic Modeling language is used to operationally define how processes work, how factors contributing to those processes interact, and what the effect of those interactions is likely to be. The framework utilizes a small set of very simple, consistent, and intuitive structural icons that can be used to represent any type of dynamic system. That structural diagram is then defined with data and sets of interrelationships which determine the expected behavior of the system over time. The history of this technique goes back to the Massachusetts Institute of Technology (MIT) in the early 1960s (1) and subsequent practitioners have expanded both the scope and rigor of this methodology. Later chapters will detail how the approach can be applied to issues in the pharmaceutical industry. However, before doing so, we need to establish the basics of the Dynamic Modeling framework and associated language.

DYNAMIC MODELING 101

The Dynamic Modeling methodology and the language which supports it is disarmingly simple yet often profound. The principle behind the approach is that any system can be represented by three conceptual elements:

1. The places where items accumulate.
2. The movement of items between those accumulations over time.
3. The information that dictates how rapidly those movements take place.

In other words, systems are composed of where things are, where those things come from and go to, and the information that determines how quickly those actions take place. These pieces constitute the *structure* of a system—the arrangement of and relationships between various system components. When these pieces are populated with data and set in motion, a dynamic model can show the resulting *behavior* of the system—how those components change over time. By explicitly capturing cause-and-effect system relationships, the Dynamic Modeling language provides a framework for both conceptualizing and analyzing the drivers and magnitude of change in a system.*

Most people are familiar with the concepts behind the Dynamic Modeling approach, as they are at the heart of how all types of systems fundamentally operate, but have not been formally exposed to the methodology. Using the Dynamic Modeling language takes some practice, and we have found that examples from a variety of situations often help emphasize the focus of the approach and introduce the associated language. Similar to translating word problems in your high school algebra or geometry class, these examples will serve as useful background and

* An excellent introductory guide to the language of Dynamic Modeling and how it can be applied to a number of real-world situations accompanies the *iThink*® software package.

Table 1 Example Monthly Checking Account Balance

Beginning balance	$10,000
Withdrawals	$5000
ATM	$500
Checks	$4500
Mortgage payment	$2500
Car payment	$500
Discretionary spending	$1500
Deposits	$5000
ATM deposit	$2000
Direct deposit	$3000
Interest earned	$10
Ending balance	$10,010

introduce the Dynamic Modeling principles that can then be applied to pharmaceutical markets.

Example 1: Account Balances*

You may have received a recent bank statement detailing the balances and transactions associated with an interest-bearing checking account, for example. Such statements are often formatted as shown in Table 1.

Table 1 clearly shows that the balance in the account is, in fact, a place where material (dollars in this case) accumulates. The balance at the end of the month is simply the balance at the beginning of the month ($10,000) plus any deposits ($5000) and interest ($10) minus any withdrawals ($5000). Those transactions are simply the means through which money has moved into or out of the account over a period of time—a month in this case.

Bank statements are very easy to convert into the Dynamic Modeling language, which provides a nice visual framework to represent financial transactions and associated metrics. First, we identify accumulations that exist at a single time point. In this case, we have only one such item; the balance in the account. In Dynamic Modeling language, such accumulations are called **stocks** and are represented with rectangles. The stock for this simple checking account example is shown in Figure 1.

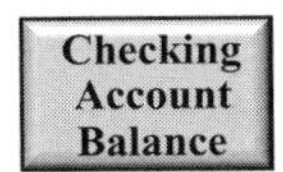

Figure 1 Stock representation of checking account balance.

* This example is adapted and extended from Kim Warren's excellent work, *Competitive Strategy Dynamics*, with his permission (3).

Stocks are time invariant in that they can be identified and counted at any single time point. In other words, if you took a picture of a system (i.e., stopped time for an instant), only the stocks of the system would show up in the photograph. Take a look around you, wherever you happen to be reading this. The books on the wall, the papers on your desk, the chairs in the airport lounge, the people in line at the bank—all would be represented by stocks. Stocks represent the current state of "how things are" and are generally easy to quantify. The number of books on your bookshelf, for example, can simply be categorized and counted at any particular point in time.

Stocks by themselves are a bit boring, as they just depict a state of being. They can be counted or quantified; but they do not "do anything" by themselves. In order for accumulations to exist and ultimately change, **flows** are needed to bring items into or out of these stocks. Flows are the dynamic movements of items that transport material (books, papers, chairs, people, money) from one stock to another. Flows are time dependent in that they exist over a period of time, and they represent "how things are changing." Stop the action in a system and the flows cease to exist. Enable the action and flows provide the means through which stocks increase or decrease over time.* Flows are always expressed as units per measure of time—salary per year, calories per day, kilowatts per month. In each case, the "per time" component indicates a flow concept. Flows are always attached to stocks, as we show by expanding our example.

From our example account statement, we can see there are three broad categories of money movements that can change the Checking Account Balance; *withdrawals*, *deposits*, and *interest earned*, and are reflected in Figure 2.

A note on iconography will help to clarify Figure 2. First, note the blackened "X" on each flow that looks something like an hourglass. These icons are intended to represent "valve regulators" much like the hot and cold knobs on older bathroom faucets. These regulators allow the flow to be turned up (increase the volume) or down (decrease the volume) and determine the magnitude of items flowing into or out of a stock over a period of time. Second, the clouds in the diagram represent the boundary of concern for the dynamic model. We know, for example, that the money coming into the Checking Account Balance through the *deposits* flow has to come from somewhere (most likely the bank account of some person or company) but since we are primarily concerned with our own balances, that stock is beyond the scope of this simple model. Similarly, the money moving through the *withdrawals* flow does not just disappear (readers with teenagers might disagree!) but rather goes into a person's wallet, purse, or cash drawer. Such dynamics,

* The stock/flow distinction was readily apparent in August 2005, as television reports showed satellite photos of the current position of Hurricane Katrina and then displayed its projected path as the weather map was "set in motion." Current positions of such storms are stock concepts as they can be captured at a single time point in still photographs. Flows come into play as the atmospheric and meteorological forces change the position of the storm over time. As was evident in 2005, the speed at which these weather dynamics unfold depends on the relative impact of these factors on the hurricane system—the devastating results of which are often hard to predict.

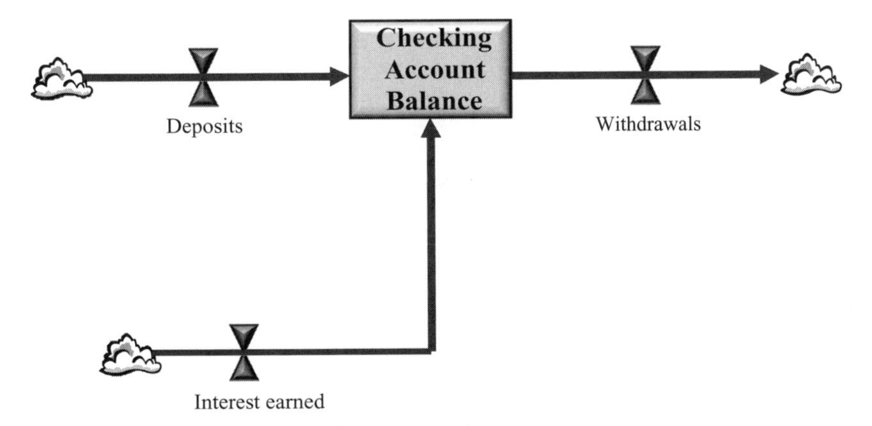

Figure 2 Stocks and flows in example checking account balance.

however, are not of concern when representing the simple bank statement. In this manner, clouds represent either infinite sources of materials or infinite sinks into which materials are deposited and no longer tracked in the dynamic model.

A fundamental principle of Dynamic Modeling is that *stocks exist because of relative flow imbalances over time.* Stocks are accumulations and exist *only* when cumulative entries (inflows) are greater than cumulative exits (outflows). As such, stocks represent "buffers" that accumulate the differences between flow rates over time. Stocks of patients on a particular drug result from a net gain of patients over the lifetime of the drug. Disease prevalence stocks exist due to the fact that more people have become afflicted with a disease than have died/recovered from it.* And stocks of prescribing doctors come about when more physicians have adopted a particular treatment than have discontinued using it. As we will show in later chapters, these various stock/flow concepts play a key role in pharmaceutical marketplaces.

With the major stock and associated flows determined, the model can now be expanded to include the information that allows the magnitude of the flows to be determined. From our sample bank statement, we know there are constituent components that determine the aggregate flow rates, as shown in Figure 3.

These constituent components in this case represent inputs (*ATM deposit* = $2000) or calculations (*checks* = *mortgage payment* + *car payment* + *discretionary spending*) and are represented in the Dynamic Modeling language as circles. Such variables are called **auxiliaries**, and are places where constants are stored, where inputs are housed, where dependent relationships between variables

* Think of the dynamics involved with infectious disease, such as the Spanish Flu epidemic of 1918–1920. For a time, the prevalence of the disease increased as the stock of people suffering from it was on the rise. But the high mortality of the disease soon meant more people were dying from it than were contracting it and this flow imbalance eventually pushed prevalence back down to where it (thankfully) stands now—zero.

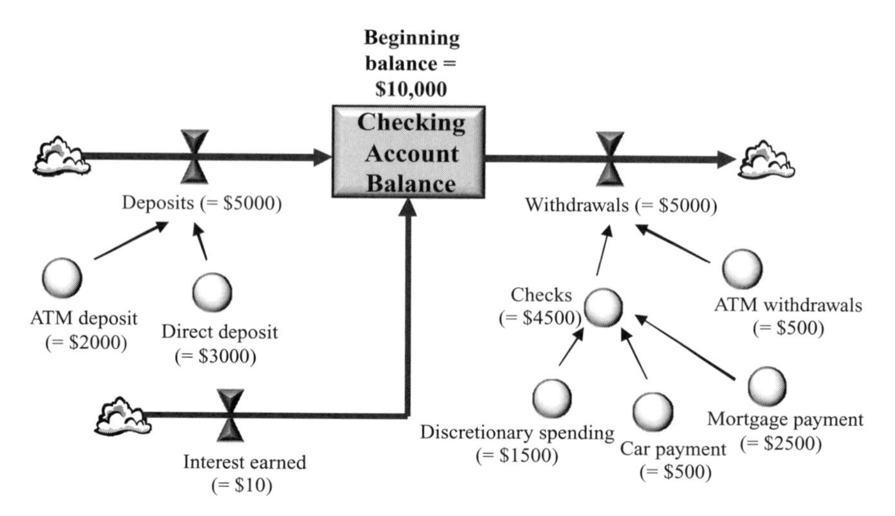

Figure 3 Stocks, flows, and associated detail in example checking account balance.

are expressed, or where calculations are made. Auxiliaries are often thought of as cells in a spreadsheet, and are useful additions to stock/flow structure. The thin arrows linking auxiliaries to each other or to flows are called **connectors** and allow for information to be exchanged between variables. Hence, the model must have connectors from the variables *mortgage payment*, *car payment*, and *discretionary spending* into the *checks* auxiliary in order for those values to be used in the equation that determines the value of *checks*. In this manner, the stock/flow diagram is augmented with auxiliaries and connectors to become the structural representation of the relationships between variables in the system, which when populated with data can calculate how those variables will change over time.

A further extension of the model involves computing *interest earned* as a function of the Checking Account Balance and an interest rate, instead of simply being a constant $10 per month. To do so, we can back-calculate or read from the account balance statement the effective monthly interest rate on the Checking Account Balance. Hence, the *monthly interest rate* is 0.1% per month ($10,000/ $10). We can include this calculation with a simple extension of the model as shown in Figure 4, which shows how *interest earned* becomes a dynamic calculation each month—Checking Account Balance multiplied by *monthly interest rate*.

In Figure 4, the connector from Checking Account Balance to the *interest earned* flow simply shows that the inflow is a function of, or depends on, the stock itself. Dynamic Modeling characterizes these types of self-referential or circular relationships as **feedback** consisting of one of two types—reinforcing or balancing. In general, feedback relates to a series of interrelationships whereby a particular variable affects other model components that ultimately "feed back" to further change the original variable. In interest-bearing checking accounts such as shown in Figure 4, the feedback relationship is very simple; *interest earned* is

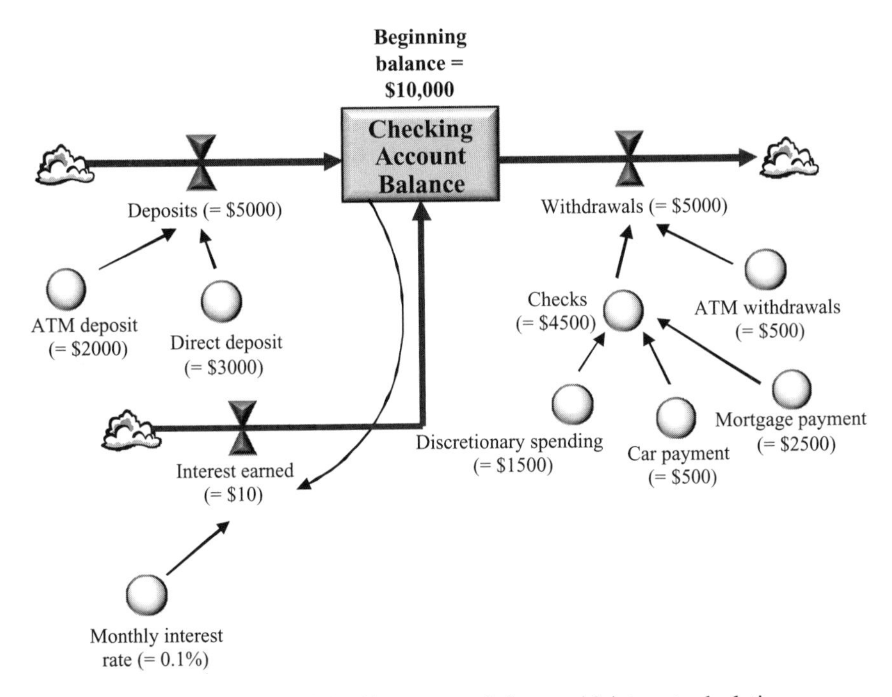

Figure 4 Dynamic model of checking account balance with interest calculation.

calculated based on the amount of the Checking Account Balance, which serves to further increase the magnitude of the stock to which it is an inflow. **Reinforcing feedback** relationships are characterized by ever-increasing change in a single direction.* Run a mental simulation to see why compound interest exhibits reinforcing feedback behavior. The greater the Checking Account Balance, the higher the inflow of *interest earned*, which pushes the value of funds in the account even higher. Reinforcing feedback relationships exhibit exponential growth, and tapping into them can be a powerful strategic option. The Dynamic Modeling framework provides an excellent methodology for conceptualizing and quantifying such circular relationships—something often missing in static or spreadsheet analyses. Reinforcing feedback relationships will be explored in more detail in chapters 8, 12, and 14, when we investigate word-of-mouth dynamics and network effects between prescribing physicians in pharmaceutical marketplaces.

Balancing feedback, in contrast, occurs when a change in a model variable creates compensating reactions which come back to change the original variable in the opposite direction. For most of us, *discretionary spending* is not a set amount each month but instead depends on how much money exists in the Checking Account Balance. There might be a minimum amount of expenses regardless of the

* For an entertaining and nontechnical look at reinforcing feedback in a variety of real world situations, see Gladwells's Tipping Point: How Little Things Can Make A Big Difference (4).

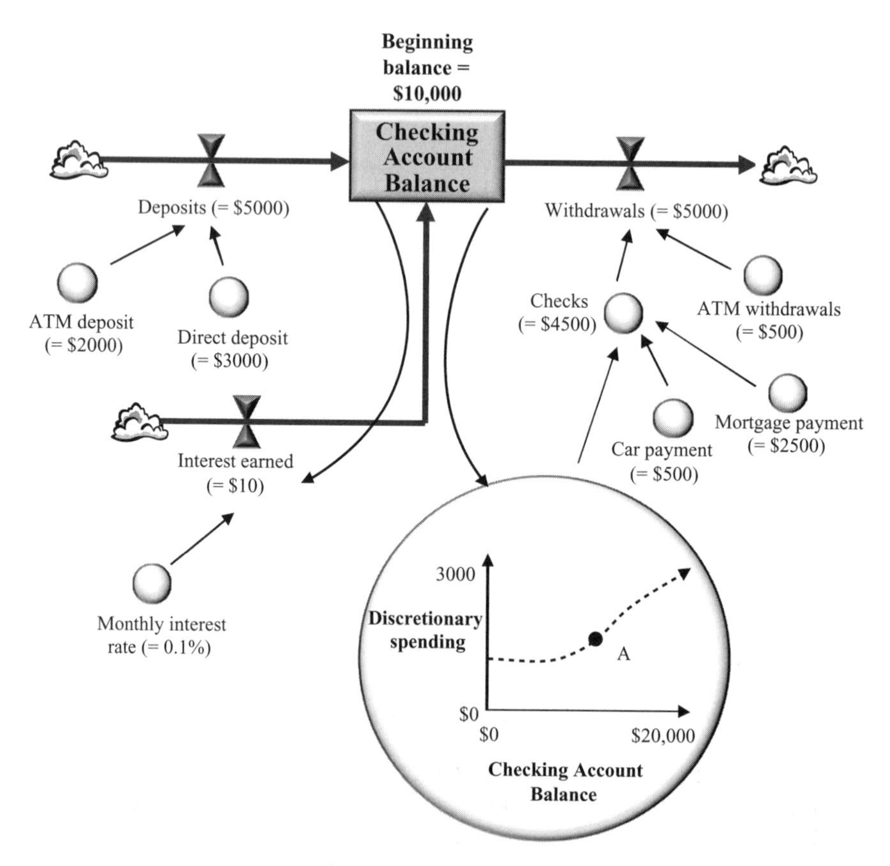

Figure 5 Dynamic model of checking account balance with interest calculation and nonlinear relationship between model variables.

value in the account, but most of us would see our *discretionary spending* skyrocket if we hit the lottery and suddenly had large amounts of funds at our disposal. This relationship between account balances and corresponding *discretionary spending* can be set up very easily using Dynamic Modeling software, including the ability to graphically depict nonlinear relationships. Figure 5 shows both the operational link between the two variables (the connector running from Checking Account Balance to the *credit card* auxiliary) and an example of a nonlinear curve relating *discretionary spending* to available funds.

Point A in Figure 5 shows the established position on the curve based on the numbers in Table 1—monthly *discretionary spending* is about $1500 when the Checking Account Balance is at $10,000. A changing balance in the account, however, due to changing magnitudes of *deposits*, *withdrawals*, and/or *interest earned* would automatically result in a corresponding change in the *discretionary spending* calculation, according to the nonlinear relationship defined. Housing such relationships in an auxiliary variable is a useful way to utilize the power of Dynamic Modeling simulation technology.

Balancing feedback mechanisms allow systems to remain in control, and are the "governors" of runaway behavior. Run a mental simulation on the relationship between Checking Account Balance and *discretionary spending* to see why this is true. The higher the value of funds in the account, the greater the amount of *discretionary spending*, which serves to bring the account balance back down by increasing the *withdrawals* outflow. Conversely, low bank account balances generally force people to tighten their financial belts, constricting the amount of *discretionary spending* and hence the outflow of *withdrawals*, thus helping to push the account balance back up. In this manner, the dynamic formulation of *discretionary spending* establishes a balancing feedback mechanism to control the Checking Account Balance.

Dynamic models such as this checking account example provide some interesting analytics and metrics which can be difficult to determine on paper or with a static analysis.

1. *The computing interval for a dynamic model can be set to any degree of specificity.* In the simple example just completed, the model calculations were established to mirror those of the checking account statement in Table 1. But suppose you wanted to look at your Checking Account Balance on a weekly or even daily basis? In most checking accounts, the interest rate is applied to the average daily balance in the account, making the arithmetic to calculate *interest earned* a difficult proposition. In addition, house and automobile payments are not spread over the course of the month, but rather are generally due on a specific time each billing cycle. Dynamic models are simulation-based; hence, the modeler can dictate the timeframe over which the relevant metrics are calculated. If you want to see your Checking Account Balance for each day over course of the month, simply set the calculation interval in the dynamic model to one day instead of one month. The model will then make 30 calculations as it simulates through the number of days in a month. Outputs can still be shown as a monthly aggregate or on a daily basis. Note that in the world of compound interest, or *any* system which does not evolve in a straight-line fashion, behavior over the course of a given month does *not* progress linearly from Day 1 to Day 30 or any points in between. Dynamic Modeling allows the choice of timeframe calculations to be easily set and/or changed, allowing for finer intervals of analysis when appropriate.
2. *Dynamic models have the ability to quickly establish nonlinear relationships between model variables, the effects of which can be easily tested through simulation.* Complicated equations can be written to describe such relationships, but Dynamic Modeling software has the ability to visually depict these mathematic relationships in a very simple and intuitive manner. Clarity is key in the business world, so effective dissemination of complex relationships often relies on an accessible communication framework.

3. *Dynamic Modeling incorporates reinforcing and balancing feedback relationships which are essential for operationally depicting the complex interrelationships driving system behavior.* The calculation methodology of Dynamic Modeling utilizes an explicit stock/flow paradigm and thereby avoids the dreaded "circular reference" errors found in a typical spreadsheet approach. The visual framework is also useful to operationally define, understand, and quantify how these feedback mechanisms actually work.
4. *The Dynamic Modeling approach involves calculation of the average amount of time items spend in a particular stock before exiting to another part of the system.* For example, suppose you wanted to know how long your money spent in your checking account before it was spent? This metric is called **average residence time**, provides a nice sanity check when investigating the relative magnitudes of stocks and flows in a truly dynamic system.

Example 2: College Student Populations

As is true with much of the Dynamic Modeling framework, you are likely familiar with the concept of residence time and may even be able to derive it implicitly without knowing the principles behind the calculation. There is value, however, in explicitly understanding this important concept, as it is both a useful mental and analytical skill when delving into the world of dynamics.

For example, imagine a four-year college or university with a student body of 4000. "How many students, on average, graduate each year?" We have asked this question to groups ranging from technical analysts to CEOs to high school students, and 90% give the same correct answer: "1000 students will graduate each year, on average." Most people make this calculation in their heads, using the logic shown in Equation 3.1.

Equation 3.1: Calculation for Determining Outflow from a Stock

Students (4000) ÷ Average time spent in college (4 years) = Students graduating/year (1000)

This relationship holds true for *any* stock that is in relative equilibrium, meaning that its value does not change significantly from one time period to the next.* In the world of Dynamic Modeling, such systems are said to be in **steady state**—the magnitude of the stocks are not changing significantly over time. Systems in relative steady state such as our checking account example or this hypothetical college student population can be examined using Equation 3.1, making it fairly easy to determine how many students would graduate under various alternate conditions; a larger university, if it takes students more than

* A stock whose magnitude is changing dramatically is more difficult to analyze in terms of average residence times. In general, information on the rate of change of the stock as well as time-series data on inflow and outflow patterns is necessary to calculate such metrics [see Ref. (5)].

Figure 6 Example stock/flow model of college student population.

4 years to graduate, or at a college which offers only two-year programs. *But this calculation can be rearranged to determine the average time spent in a stock if only the stocks and flows are known.* For example, assume we have established the stock/flow model and associated data as shown in Figure 6.

In this case, rearranging Equation 3.1 with the data from Figure 6 is shown in Equation 3.2.

Equation 3.2: Determining Average Residence Time in a Steady-State Stock

Students (5000) $\div$ Students graduating per year (1000) $=$ Average time spent in college (5 years)

By utilizing the equation for steady-state accumulations, the relative magnitudes of a stock and its associated flow(s) can be used to determine how long, on average, items (college students, dollars, etc.) have spent in a particular stock.* This metric becomes a useful way to "sanity-check" the results of a dynamic model in order to determine if the calculated residence times in various stocks in the system seem reasonable. Returning to our bank statement example, the calculation for residence time suggests that deposited funds stay in the checking account for two months, on average, before being spent—Checking Account Balance of $10,000/*withdrawals* of $5000 per month $=$ 2 month average residence time. As we will show in subsequent chapters, the analysis of residence times can often elicit aspects to dynamic behavior which might not otherwise be readily apparent.

At this point, some readers might question the advantage of setting up a structural, stock/flow model instead of a simple spreadsheet. The examples in this chapter could have been completed by hand or on the back of an envelope, without the use any confusing "boxes and pipes." But in our 75 collective years of consulting experience, both in the pharmaceutical industry and elsewhere, we have found that *the Dynamic Modeling language is the absolute best way to conceptualize, communicate, and ultimately analyze the operational processes and relationships that determine how systems change over time.* Other publications explore the benefits of the Dynamic Modeling approach in more depth, and are excellent

* For example, observers of local housing markets often see data regarding how long, on average, homes stay on the market before being sold. This metric is always calculated using this stock/flow principle—taking the number of homes for sale and dividing it by the number of homes sold in the past month.

references for readers interested in the details of this methodology. The rest of this work will detail the Dynamic Modeling approach as applied to pharmaceutical markets, and show how this framework can provide insight into market analysis, strategy development, and the interrelationships that ultimately determine how these markets behave over time. But in short, the Dynamic Modeling approach and associated language can add value in a number of ways.

ADVANTAGES OF THE DYNAMIC MODELING APPROACH

Dynamic Modeling can be beneficial to a variety of analytic situations for the following reasons:

1. Stock/flow models set up explicit, easy-to-read structure in a rigorous framework.
2. Dynamic Modeling framework is a useful means to convey ideas, compare mental models, and communicate strategic insights in a format which is not subject to the inconsistencies of written or spoken language.
3. Stocks and flows are a fundamental aspect of *any* system, and are thus easily conceptualized even by people with nontechnical backgrounds.
4. Dynamic Modeling methodology operationalizes dynamic concepts, allowing teams to explicitly define the sets of cause-and-effect relationships driving subsequent behavior.
5. Stock/flow models can be simulated to show the implications of their associated data assumptions over time.
6. Dynamics models provide a tool to assemble, integrate, and validate various input data, ensuring a consistency of assumptions that can be tested via simulation.
7. Outputs from dynamic models can be difficult to ascertain using alternative approaches, as static analytical tools are often not designed with such dynamics in mind.
8. Dynamic simulations are fully adjustable with respect to the computing intervals, allowing outputs for both the short term (daily or weekly) and medium term (monthly or quarterly) [and sometimes even long term (yearly)] within the same model.

The benefits of the Dynamic Modeling approach will be evident as we progress through the pharmaceutical examples in rest of this work. Our third example in this chapter, however, will provide one last look at simple stock/flow structure to explicitly capture dynamic behavior in an area familiar to most readers; changing age demographics in a defined population.

Example 3: Aging Chains

Demographic projections often include an age segmentation dividing a population into specific age categories. These segmentations usually break up a constituent

Table 2 The U.S. Population Forecast by Age, 2000–2050 (In Thousands)

Age group	2000	2010	2020	2030	2040	2050
0–19	80,549	83,236	88,887	95,104	101,625	109,147
20–64	166,515	185,456	192,285	197,027	210,270	224,001
65+	35,061	40,243	54,632	71,453	80,049	86,705
TOTAL	282,125	308,935	335,804	363,584	391,944	419,853

Source: U.S. Census Bureau, 2004, "U.S. Interim Projections by Age, Sex, and Hispanic Origin," http://www.census.gov/ipc/www/usinterimproj/, Internet release date: March 18, 2004. (Rounding error may be present.)

population into age groups, the distribution of which may or may not change over a specified interval. For example, the U.S. Census Bureau publishes population demographics forecasts, as shown in Table 2.

The mathematic methodology behind these types of projections is often quite complicated, but these types of demographics forecasts often have huge impact when evaluating market potential of many products, including pharmaceuticals. The forecast interval in this case is beyond even the long-term planning horizon in the pharmaceuticals industry, but should provide an illustrative example to examine the application of the Dynamic Modeling approach to such situations.

While instructive, simple static projections do not explicitly address the inherent dynamic structure that operationally describes what causes these demographic shifts over time. However, a very simple dynamic model cannot only produce these results but also provide additional insight into demographic dynamics, as shown in Figure 7.*

The advantages of having both the stock/flow structure of the system and the resulting age distribution behavior over time are numerous.

1. The Patients Age 0–19 and Patients Age 20–64 stocks have inherent time components associated with them, such that patients stay in each stock for 20 and 45 years, respectively, before *aging* and advancing to the next age group.

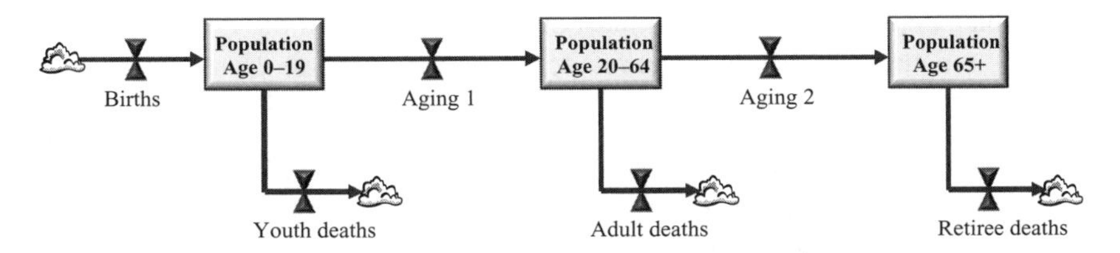

Figure 7 Example stock/flow structure of population dynamics.

* For simplicity, this structure ignores net population migration, but can be included as an easy extension.

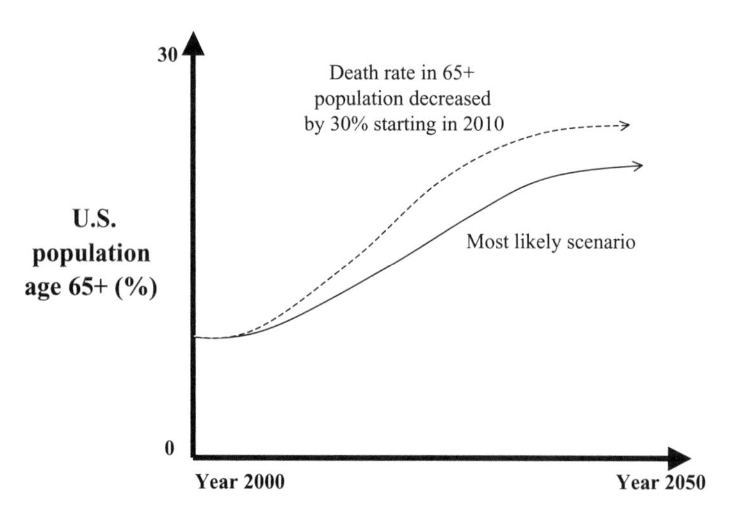

Figure 8 Example outputs of dynamic model of age demographics—base case vs. decreased death rate scenario.

2. The Patients Age 65+ does not need an aging mechanism, as patients will stay in that stock for some average length of time until they exit through the *retiree deaths* flow.
3. Age-specific death rates can be applied in order to determine the values of the *youth deaths*, *adult deaths*, and *retiree deaths* flows.
4. The dynamics of the aging chain allow for a finer level of analysis. For example, the flows of *aging 1* and *aging 2* can be computed to show the gain or loss of patients from different age categories over time.
5. The explicit stock/flow structure and resulting behavior allows for simple but important scenario testing. For example, the static demographic forecast would not show how changing age-specific death rates (resulting from cures for various diseases and/or improving health care access, for example) would affect population numbers and distributions. However, a dynamic model of population demographics easily calculates such metrics, as shown in Figure 8, which depicts the projected impact on the distribution of the Population Age 65+ in the event that the death rate for that population was decreased by 10% starting in 2010.

Aging chains combine the simplicity of the Dynamic Modeling language with the power of simulation technology to produce true operational dynamics over time. Such chains can be adapted to include any number of age categories, or disaggregated for various market segments (gender, race, geography—see chap. 5). In fact, these types of dynamic models are often used to estimate the effects of changing demographics on government services. For example, the aging of the baby boomer generation, combined with increased life expectancy, will likely put increasing demands on the U.S. Health Care system and Social Security/Medicare safety nets. Such dynamics, which play out over long periods of time and are not strictly linear, are difficult to capture in static models or straight-line projections.

The three basic examples in this chapter show how simple stock/flow diagrams can be extended, modified, and adapted to address increasingly complicated situations or problems. Combining these diagrams with Dynamic Modeling simulation technology provides a powerful tool for analysis and understanding of the behavior of complex systems. Even very detailed dynamic models, however, start with a very aggregate, 10,000-foot view and rely on the four basic building blocks (stocks, flows, connectors, and auxiliaries) to describe the structure and resulting dynamics within the system. More complete descriptions of the components and complexities of the Dynamic Modeling language can be found in other works that focus on the details of this methodology. For our purposes, this introduction allows us to turn our attention to the use of the Dynamic Modeling approach in specific pharmaceutical market situations.

The Value of Stock and Flow Thinking

Steve Peterson
Principal
Lexidyne, LLC

Since the early 1980s, I have been using the stock/flow framework as a vehicle to represent and simulate dynamics in a wide range of issue areas. During that time, I have had multiple opportunities to use stocks and flows in very practical ways to help individuals, teams, and organizations build understanding and improve performance. From my perspective, the stock/flow framework adds value in four distinct ways:

1. Because the framework is simple, generic, and very physical in its orientation, *anyone* can learn to converse in the language of stocks and flows. The language can become an "organizational Esperanto" of sorts, enabling cross-functional teams to transcend their functional stovepipes as they communicate in a common language. As a result, the language facilitates the process of quickly getting people on the same page around a shared objective.
2. Because it provides a clear operational distinction between *accumulations* and the *activities* that cause accumulations to grow or decline over time, the stock/flow framework makes it possible systematically to identify levers for improving or otherwise changing performance. For *any* accumulation, the pathways for changing its magnitude are found among its associated flows. A stock/flow map therefore can become the vehicle for identifying potential intervention points within a system or process. Further, the diagram can become the focal point for identifying the delays and process inertias that might impede the effectiveness of a particular intervention strategy.

3. Because the stock/flow framework supports mapping of physical accumulations and flows as well representation of information flows, feedback linkages, and business rules that drive the specific magnitude of flows, it facilitates rigorous operational thinking about the full range of impacts of a proposed course of action. In a market situation, for example, a stock/flow model can be used to get a better handle on the full range of consequences—intended and unintended, short-term and long-term—of a particular strategic option.
4. Because the stock/flow map is a visual representation that lies above a sophisticated mathematical model, the framework can facilitate data collection and analysis activities. For the results of a dynamic model to make sense, the input data to a model must be internally consistent and validated through simulation. The process of arriving at an integrated set of internally consistent inputs to the model typically identifies key data gaps and helps to resolve the issues that arise when different data sources have divergent numbers.

Regardless of the industry or function, client teams tend to relate well to the structured approach that is provided by the stock/flow framework. In my experience, the Dynamic Modeling methodology is an excellent vehicle, both for coming to a common understanding of the mechanisms that drive the behavior of a dynamic system and for designing initiatives for improving that behavior.

SUMMARY

- Dynamic models focus on the structure and interrelationships which cause systems to change over time using an operational language to describe dynamics.
- All dynamic models are built using four basic building blocks; stocks, flows, connectors, and auxiliaries.
- These building blocks can be used to develop a dynamic model for any type of system.
- Stocks represent accumulations of items that can be counted at a single time point.
- Flows represent the dynamic movement of items over time that ultimate change the magnitude of the stocks to which they are attached.
- Connectors allow information to be transferred between model components in order to define the magnitude of flows or to make calculations relevant to model dynamics.
- Such calculations are housed in auxiliaries, which also serve as storage cells for constants or as inputs to flows or other auxiliaries.
- Reinforcing feedback relationships produce ever-increasing change in one direction, and represent potential for exponential growth behavior.

- Balancing feedback relationships tend to balance themselves out, creating a series of causal relationships which circle around to maintain systems in some sort of balance.
- Average residence time relates to how long items spend, on average, before exiting.

REFERENCES

1. Forester JW. Industrial Dynamics. Cambridge, MA: MIT Press, 1961. Currently available through Pegasus Communications: Waltham, MA.
2. Richmond B, Peterson S. An Introduction to Systems Thinking. From the users manuals for the ithink software. Lebanon, NH: Available from iseesystems, Inc., 1996.
3. Warren K. Competitive Strategy Dynamics. Chichester, UK: John Wiley & Sons, 2002.
4. Gladwell M. Tipping Point: How Little Thing Can Make A Big Difference. Boston, MA: Back Bay Books, 2002.
5. Sterman J. Business Dynamics: Systems Thinking and Modeling for a Complex World. Boston, MA: McGraw Hill, 2002.

4

Patient Flow Sector of the Standard Template Dynamic Model

INTRODUCTION

This chapter will cover the following topics:

- Properties of Standard Template indications
- Incorporating epidemiology using stocks and flows
- Compliance/Persistence
- Revenue calculations
- Treatment options
- Treatment change matrices
- Patient Flow allocation

From our very first engagement in the pharmaceutical industry to the present day, our clients have asked for help evaluating the markets for drugs treating chronic illnesses. The reasons for this recurrent request are simple; chronic diseases affect roughly half of the U.S. adult population, and spending to treat such conditions represents over 70% of all health care expenditures (1,2). Aging of the Baby Boom generation, longer life spans, and the push for increased health care coverage virtually guarantee that chronic illnesses will continue to be a huge part of the pharmaceutical market landscape over the next few decades.

In response to the demand from our clients in the pharmaceutical industry, we have used the "Dynamic Modeling" methodology to develop what we call the "Standard Template"—an expandable stock/flow framework incorporating epidemiology, treatment options, and compliance/persistence metrics to evaluate and analyze patient behavior in individual indication marketplaces. Later chapters will address the extension and expansion of this basic methodology, incorporating aspects of physician choice and treatment attractiveness of therapy options. But even by itself the Standard Template is an extremely useful framework for analyzing how patients behave in pharmaceutical markets, and disease indications which fit nicely within the template generally have three common properties.

PROPERTIES OF DISEASE INDICATIONS FITTING INTO THE STANDARD TEMPLATE DYNAMIC MODEL

1. *Chronic disease* = Lifetime diseases with no cures, only treatments.
2. *Defined pharmaceutical treatment options* = Data is available on what types of treatment patients receive.
3. *Patient churn* = Patients often discontinue, reinitiate, and/or switch treatments during the course of therapy.

The list of diseases meeting these criteria is extensive and often high-profile, including hypertension, diabetes, congestive heart failure, Alzheimer disease rheumatoid arthritis, mental disorders such as schizophrenia, bipolar disease, and depression, dislipidemia, and viral infections such as hepatitis and HIV. Each of these markets has unique characteristics, but the basic approach of the Standard Template can capture many key strategic dynamics and can then be customized to the specifics of the individual indication.

EPIDEMIOLOGY

The Standard Template first details full epidemiology including a clear distinction between prevalence and incidence. In the stock/flow language, the former represents where patients exist *at a single point in time*, and the concepts of prevalence are therefore shown as stocks as seen in Figure 1.

Prevalence is expressed in two ways; as a percentage of the population that has a particular disease at a point in time, or as an absolute number of people with that disease. For example, a report by the Centers for Disease Control

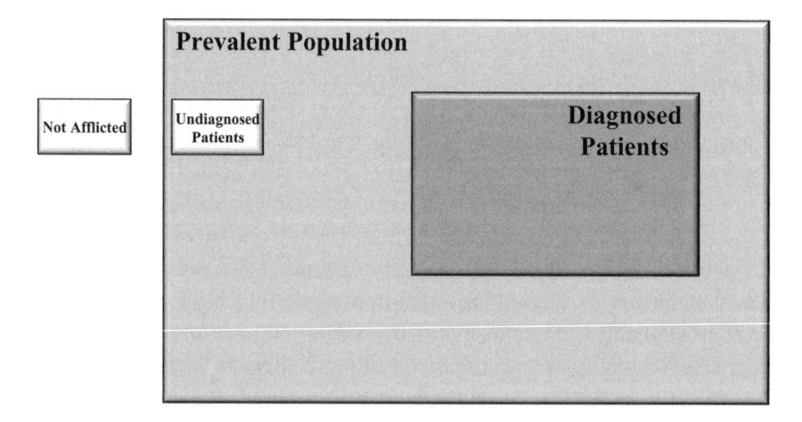

Figure 1 Stocks representing key epidemiology concepts.

and Prevention says that 18.2 million U.S. citizens had diabetes in 2002 (3)—a number easily converted to percentage terms by comparing to published population statistics. This type of prevalence estimate determines the initial number of patients in the Prevalent Population stock. Readers familiar with epidemiology will note there are actually a variety of prevalence metrics for various diseases. For the purposes of this analysis, we are using point prevalence; the number of people with a particular affliction at a single time point. Period or lifetime prevalence formulations are an easy extension (see chap. 5).

The concept of point prevalence can be further divided according to prevailing diagnosis rates, which places patients in the Undiagnosed and Diagnosed categories. Projections of disease epidemiology vary by methodology and result, so some triangulation may be necessary to understand differences in epidemiology forecasts. Some published research uses sophisticated trend analysis to project changes in demographics segments such as age, race, and gender. Prevalence rates for each group are then applied to projected population statistics to arrive at an integrated forecast of disease prevalence. At times, parameter estimates from a number of sources must be pieced together to develop a full epidemiology table at a series of future time point, such as is shown in Table 1.

A distinct advantage of a dynamic model is the ability to apply forecasted values to a stock/flow framework, which when coupled with mortality rate information allows the associated epidemiology flows to be determined. To do so, the Standard Template fills out the basic epidemiology picture as shown in Figure 2.

The inflow to the stock of Prevalent Population is, by definition, *true incidence*—the number of patients that develop a particular disease over a period of time, usually a year. In chronic diseases, a patient usually develops a disease physiologically before being diagnosed by the medical community; so *true incidence* is an inflow to the stock of Undiagnosed Patients. Two things can subsequently happen to these patients after some period of time: (*i*) They can die before being diagnosed, as is represented by the first outflow of *deaths from Undiagnosed* (the cloud into which they flow means they are no longer of concern in the model) and (*ii*) they can be diagnosed with the disease and enter the stock of Diagnosed Patients through the flow labeled *diagnosed incidence*. In chronic diseases, a

Table 1 Diabetes Prevalence and Diagnosis Forecast (2000–2050)

Variable	2000	2005	2010	2025	2050
Population	275,238	287,465	299,661	338,081	403,814
Prevalence	15,375	17,466	19,802	27,878	40,761
Undiagnosed	4393	4990	5658	7965	11,646
Diagnosed	10,982	12,476	14,144	19,913	29,115
Prevalence Rate (%)	5.59	6.08	6.61	8.25	10.09
Diagnosis Rate (%)	71.4	71.4	71.4	71.4	71.4

Source: Adapted from Ref. 6.

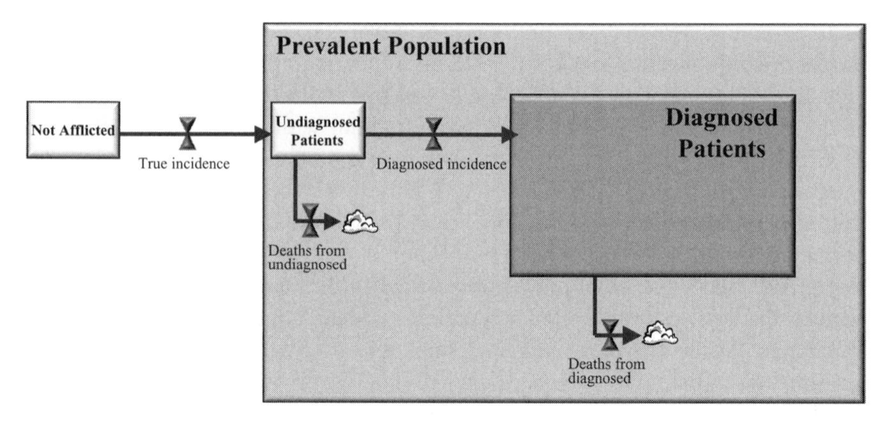

Figure 2 Stock/flow model representing key epidemiology concepts.

diagnosed patient is "red flagged" for life—once you are categorized as a diabetic, for example, you are labeled as such for your entire lifetime. Without the possibility of true recovery from the illness, the only way to exit the Diagnosed Patients stock, unfortunately, is to die and exit the system through the *deaths from Diagnosed* outflow.*

Although prevalence forecasts rarely include estimates of incidence, these very important flows can be back calculated fairly easily with disease-specific mortality data. To demonstrate, we will introduce a simplified numerical example that will be extended throughout this book as increasingly complex aspects of the Standard Template are incorporated.

The Value of Epidemiology in the Dynamic Modeling Framework

Cynthia Willey, PhD
Managing Director and Senior Epidemiologist
Episource.com

Our business began when clients in the pharmaceutical industry consulted our epidemiologists about why estimates of the prevalence of disease varied as much as 200%? We find that many prevalence estimates are promoted on the Internet, but it usually takes an experienced epidemiologist to help choose the estimate that best fits each client's needs and the specific patient flow model strategy.

* The history of incorporating stock/flow concepts to estimations regarding changing populations goes back to the 17th century, although the methods for doing so at that time were much less sophisticated [see Ref. (7)—a book on the history of risk assessment].

Basic, descriptive epidemiology is the science of documenting how diseases vary between older and younger people, males and females, or other population groups. At Episource.com we utilize nationally representative, epidemiologic studies (such as those released by the National Center for Health Statistics), to provide our clients with the most up-to-date and accurate projections of chronic and infectious disease prevalence in the United States, Europe, and Japan. In many cases, our epidemiologists analyze newly released data before published estimates are available. As more information becomes available about specific diseases, treatment, and compliance patterns, we continually update our reports.

The Standard Template incorporates epidemiology, treatment options, and compliance/persistence metrics to evaluate and analyze patient behavior in individual indication marketplaces. It allows sophisticated manipulation of data, and is a powerful way to analyze the evolution of a disease indication, but a key element is accurate epidemiologic data that is consistent with the model specification. In our experience, we have found that definitions of disease often need to be modified to include a more restrictive or broader set of patients that match the clients' intentions for marketing a product. Good communication between the epidemiologist, the Dynamic Modeling team, and the market research team is essential for developing customized patient flow models.

Our input data for patient flow models represents the experience of all patients in each country or area of interest and is a key component of the Standard Template. As epidemiologists, we are always excited about each new opportunity that contributes to the basic foundation of sophisticated dynamic models that provide remarkable insight into future disease trends and treatment options.

INTRODUCTION TO EXTENDED STANDARD TEMPLATE EXAMPLE

For illustration purposes, assume an epidemiology forecast has been developed for a chronic indication as seen in Table 2.

The Standard Template allows for calculation of the associated epidemiology flows with the help of some basic arithmetic based on mortality statistics. For

Table 2 Sample Epidemiology Stocks in the Standard Template (End of Calendar Years 2009–2015)

Variable	2009	2010	2011	2012	2013	2014	2015
Population	1000	1038	1108	1158	1198	1238	1307
Not Afflicted	600	623	665	695	719	743	785
Prevalence	400	415	443	463	479	495	523
Undiagnosed	100	105	111	116	120	124	131
Diagnosed	300	310	332	347	359	371	392

Table 3 Sample Epidemiology Stocks and Associated Mortality

Stock	Magnitude	Annual death rate (%)	Annual deaths from stock
Undiagnosed Patients	100	2.0	2
Diagnosed Patients	300	2.0	6

example, if published reports indicate that 2.0% of patients suffering from a particular disease die in an average year,* the magnitude of the death flows can be quickly determined, as shown in Table 3.

This mortality information can be integrated with the epidemiology forecast to determine the other key flows in the model, as shown in Figure 3.

From Table 2, the forecast indicates the stock of Diagnosed Patients will increase from 300 to 310 from the end of 2009 to the end of 2010. For the stock to go up by 10, knowing that 6 will die during 2010, the inflow of *diagnosed incidence* must be 16 over the course of the year. Take a moment to examine the stock/flow structure and associated parameters to see why this must be true. By creating an explicit stock/flow structure which can be populated with actual data, dynamic models ensure internal consistency by operationalizing the interrelationships between key epidemiology concepts.

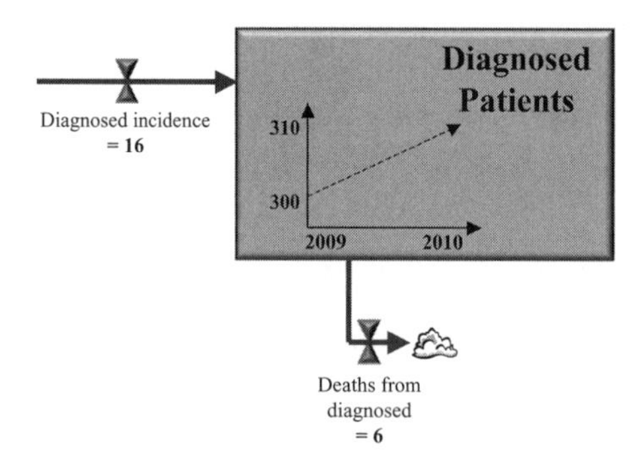

Figure 3 Stock/flow dynamics and associated sample data for diagnosed patients in the Standard Template.

* If disease-specific mortality data is not available, the overall death rate for the defined population can be used as a starting point. Mortality estimates in this form must capture death from all causes for people suffering from the disease in question. In other words, the death rate of 2.0% per year includes deaths *due* to the particular disease as well as deaths caused by other factors. For example, the Diabetes Fact Sheet from the Centers for Disease Control and Prevention estimates that the increased mortality factor for diagnosed diabetes patients is 2—patients suffering from this disease die at a rate that is twice that of the nondiabetic population. This metric can also be modified and applied to specific patient segments individually—see chapter 5.

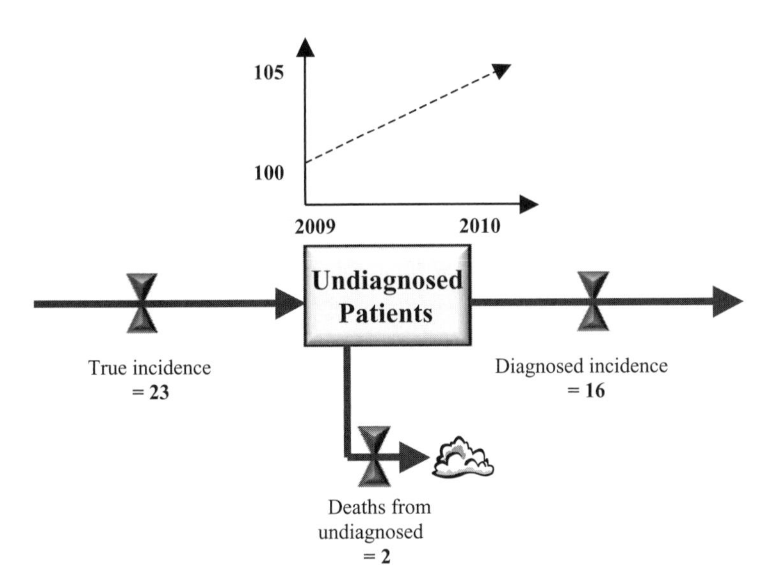

Figure 4 Stock/flow dynamics and associated sample data for undiagnosed patients in the Standard Template.

A similar calculation can be performed for the stock of Undiagnosed Patients as shown in Figure 4.

The fully populated diagram for the period starting at the end of 2009 and terminating at the end of 2010 (with stock values in **bold** and flow number in *italics*) is shown in Figure 5.

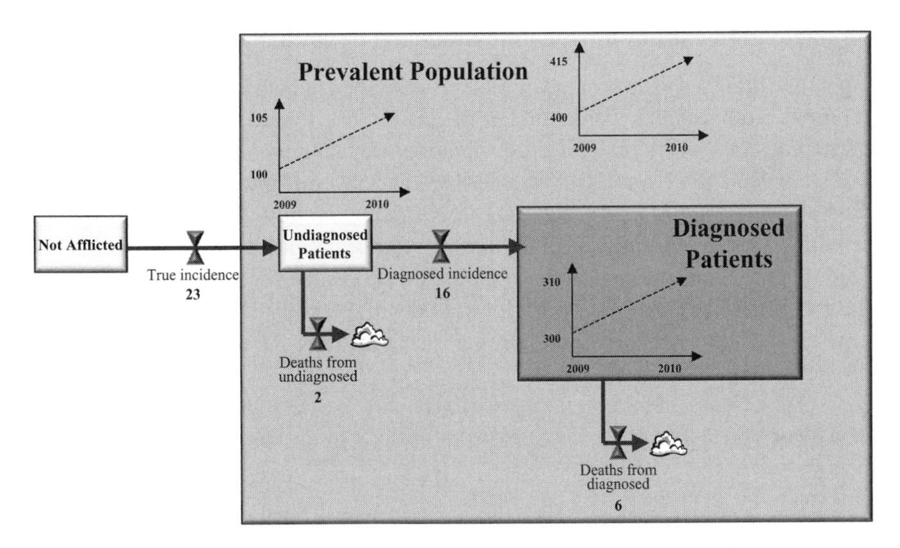

Figure 5 Stock/flow dynamics and associated sample data for the Standard Template.

Table 4 Sample Epidemiology Stocks and Flows 2009–2015

Variable	2009	2010	2011	2012	2013	2014	2015
Population	1000	1038	1108	1158	1198	1238	1307
Not Afflicted	600	623	665	695	719	743	785
Prevalence	400	415	443	463	479	495	523
Undiagnosed	100	105	111	116	120	124	131
Diagnosed	300	310	332	347	359	371	392
true incidence		23	36	29	26	26	38
diagnosed incidence		16	29	22	19	19	28
deaths 1		2	2	2	2	2	3
deaths 2		6	6	7	7	7	8

Recall that the stock numbers represent the number of patients in various accumulations at single point in time. In the example shown in Figure 5, the stock values are shown for both the end of 2009 and the end of 2010 to allow readers to see the year-to-year changes and compare them to the associated flow numbers. Later examples will just show the stock values as a single time point. Flow values, on the other hand, exist over a period of time. In this example case, the flows represent the magnitude of patient movements occurring over the specified interval—2010.

In this fairly straightforward example the associated calculations could easily be done by hand or in a spreadsheet. However, the dynamic model can integrate this process to show yearly or even monthly prevalence and incidence numbers which automatically adjust to changes in epidemiology forecasts, as seen in Table 4.

Strategic Implications of Dynamic Modeling of Epidemiology

Note that Table 4 shows the flow implications associated with epidemiology stock forecasts from Table 2. For example, the prevalence forecast does not appear to be unrealistic, but the associated calculations for *true incidence* imply a series of "*spikes*" in 2011 and 2015 in order to meet overall prevalence numbers. A similar situation exists with the *diagnosed incidence* flow. Stock/flow structure provides a more accurate and enlightening representation of chronic disease epidemiology which can be used as a sanity check against forecasts of prevalence, diagnosis rates, and key incidence flows. Many of our client teams have used these types of Dynamic Modeling structures to assess the reasonability of epidemiology projections.*

The *Dynamic Modeling* framework can now be expanded to include aspects of patient flow between various treatment options in the disease indication. This

* Parameter estimates, particularly those in epidemiology forecasts, often have ranges associated with them. Our extended example will show deterministic data estimates, but uncertainty regarding input parameters is easily incorporated using Dynamic Modeling software. See chapters 14 and 16.

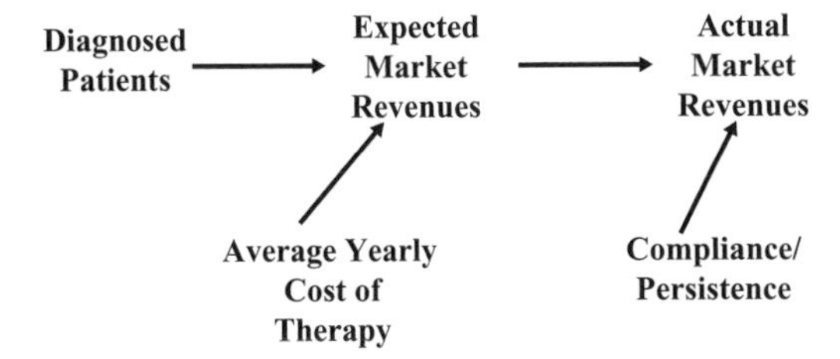

Figure 6 Common model of compliance/persistency metrics in pharmaceutical market evaluations.

integration of epidemiology and marketplace dynamics make the Standard Template a powerful way to analyze the evolution of a disease indication and better understand the patient dynamics at play within it.

COMPLIANCE/PERSISTENCE

Compliance and persistence are frequently used terms in the pharmaceutical industry, sometimes interchangeably and often inconsistently. In general, these words are intended to imply the "falloff" which occurs when attempting to translate epidemiology into revenue dollars. At their most basic and static models of compliance/persistency are often represented as shown in Figure 6.

The framework shown in Figure 6 is not technically wrong, but masks some important distinctions between compliance and persistency that might be important from a strategic standpoint. A combined compliance/persistency term represents patients either discontinuing medication or not taking their prescribed number of pills. There is a large dynamic and psychological distinction between these two concepts, however, and the Standard Template depicts this difference in a very simple operational form.

Stock/flow structure depends on precise definitions, and the vagueness surrounding an aggregate compliance/persistency term requires clarification. In the Dynamic Modeling methodology, **compliance** is a continuous measure of number of pills a treated patient actually takes versus the amount actually prescribed. For example, a diabetic patient taking 5 pills in 10 days for a drug prescribed as a once-a-day therapy would have a 50% compliance rate. A 50% compliance rate cuts expected drug revenues in half, as prescriptions intended for monthly refill are instead refilled every 60 days. In strict terms, compliance relates to treated patients which are still revenue generating, albeit at a reduced rate. This metric varies from patient to patient, but aggregate true compliance rates can be determined from both market research and database analysis.

The term **persistence** represents the average time patients stay on a particular drug. In reality, this concept should be called drug-specific persistence, and it encompasses two different dynamics. The first dynamic involves patients

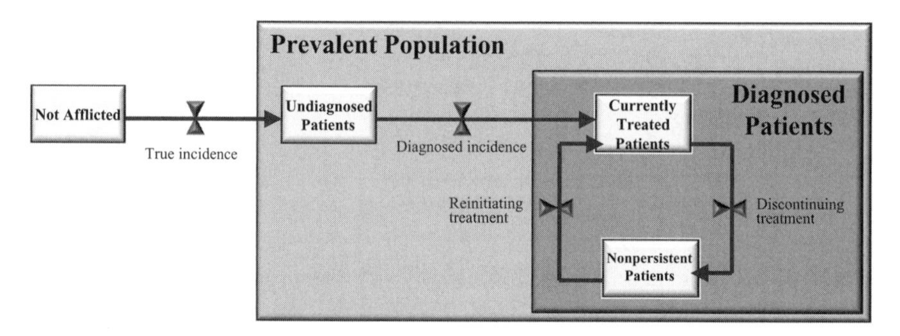

Figure 7 Standard Template including nonpersistent patients and associated flows.

switching drugs from treatment A to treatment B, for example. The second dynamic accounts for patients discontinuing therapy entirely, they stop taking Treatment A or *any* other form of medication. The Standard Template addresses these issues by introducing a concept of Non-Persistent Patients; those who are no longer taking *any* form of prescribed medication. The remaining Diagnosed Patients are deemed to be "Currently Treated," as shown in Figure 7.*

The dynamics surrounding Non-Persistent Patients are important in many chronic indications, and many companies focus on patients *discontinuing treatment*. Again, according to the precise definitions of the Standard Template, this flow captures patients stopping *all* forms of medication, not simply switching from one treatment to another. However, the flow labeled *reinitiating treatment* is often neglected. Run a mental simulation on the Standard Template structure to see why this flow is important. If patients only discontinued treatment, all else being equal, the stock of Non-Persistent Patients would increase over time. In many markets, however, the relative numbers of Currently Treated versus Non-Persistent Patients is changing only slightly, if at all. Therefore, there needs to be a pathway for patients to return to therapy after they have spent some time being Nonpersistent. In fact, for the distribution of the stocks to be fairly constant over time, the magnitude of the *discontinuing treatment* and *reinitiating treatment* flows need to be roughly equal. Depending on the relative rates of therapy switching versus discontinuation, the *reinitiating treatment* flow often represents a large opportunity for capturing patients as they return to therapy.

By definition, true compliance should apply *only* to the Currently Treated Patient stock, as the dynamic structure has explicitly accounted for the Non-Persistent Patients who are taking no medication. Auxiliary variables and connectors (see chap. 3) can now be incorporated to augment the stock/flow diagram. Recall that auxiliaries represent inputs or computations, much like cells in a spreadsheet, and connectors show relationships and information exchange between variables, shown in Figure 8.

* For simplicity, subsequent diagrams will not show death flows from various stocks, although they exist unless otherwise noted.

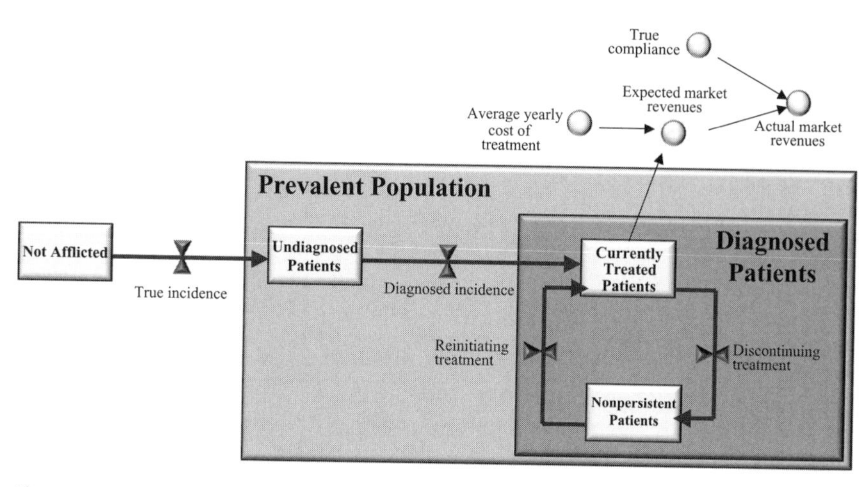

Figure 8 Market revenue calculation structure using the Standard Template.

The auxiliary variables dealing with *average yearly cost of therapy* and *true compliance* metrics allow for market revenues to be easily calculated. For simplicity, assume the stock of Currently Treated Patients is 200 and remains constant over the course of 2010. Table 5 shows the associated calculations for revenue generation in the Standard Template.

Recall that the dynamic model is not limited to yearly averages, and however, actually computes monthly revenues based on changing values in the system. For example, the dynamic simulation might produce the following results based on changing numbers of patients in the Currently Treated Stock, as seen in Table 6.

It is worth reviewing some dynamic possibilities which can occur when a simulation is computing on a monthly time frame but reporting yearly numbers. As seen in this chapter's example, the Standard Template runs on a monthly basis and allows for outputs to be aggregated into yearly numbers. This calculation versus aggregation issue is clear regarding financial metrics such as revenues, but can be a bit confusing when looking at patient dynamics. As an example, imagine a single patient named John who develops a disease at the start of a particular year. John

Table 5 Sample Yearly Revenue Calculation—Standard Template

Variable	Value/calculation
Currently Treated Patients (2010)	200
average yearly cost of therapy	$6000
expected market revenues	$1,200,000
true compliance	60%
actual market revenues (2010)	$720,000

Table 6 Sample Monthly Revenue Calculation—Standard Template

Variable	January 2010	February 2010	March 2010	April 2010	May 2010	June 2010
Currently Treated Patients	200	201	200	190	195	200
average monthly cost of therapy	$500	$500	$500	$500	$500	$500
expected monthly market revenues	$100,000	$100,500	$100,000	$95,000	$97,500	$100,000
true compliance (%)	60	60	60	60	60	60
actual monthly market revenues	$60,000	$60,300	$60,000	$57,000	$58,500	$60,000
Cumulative Revenues	$60,000	$120,300	$180,300	$237,300	$295,800	$355,800

might go through the flow of *true incidence* in January, sit in the Undiagnosed Patients stock in through April, then become a Currently Treated Patient in May following a physical at which his disease was identified and he was counted in the flow of *diagnosed incidence*. He is started on therapy in June, during which time he generates market revenues by purchasing medication. John may be diligent and refill his prescription in July, but soon he finds himself not taking his daily medication, so his 30-day script is not refilled until September. Perhaps in October he feels better, does not want to pay for any more medication, so he stops taking his pills entirely. At that point, he exits through the *discontinuing treatment* flow and becomes a Non-Persistent Patient, where he sits until December when he refills his script (*reinitiating treatment*) and returns as a Currently Treated Patient. Note in this scenario, John can be in only one of the stocks at any point in time, defined as an individual month, but he has been counted as part of each of the four flows over the course of the year. Later chapters will detail categorizing, quantifying, and analyzing such treatment changes, which are important as they represent opportunities to direct patients to drug therapy in general or a specific treatment option in particular. For now, it is important to realize that a monthly calculation interval allows for a single patient to be involved in multiple patient flows throughout a yearly interval.

Strategic Implications of Dynamic Modeling of Compliance/Persistence

By explicitly capturing the operational differences between patient compliance versus treatment persistence, the Standard Template is able to fundamentally represent these key concepts in a manner consistent with effective Brand Planning. Defining and quantifying true compliance appropriately may help identify a key strategic lever in cases where patients remain on therapy but do not take all of their daily prescribed medication. Such situations may suggest a compliance program strategy designed to track compliance of patients and enact marketing programs designed to increase this metric. Conversely, identification of low true persistency rates may suggest strategy of designed to keep patients from discontinuing treatment entirely. These types of detailed strategic insights come from operationally defining and quantifying the differences between compliance and persistency in pharmaceutical marketplaces.

TREATMENT OPTIONS

The final piece to the Standard Template structure is the disaggregation of Currently Treated Patients into various forms of treatment options, which provides the dynamic model with additional flexibility and power. This expansion

- categorizes and tracks patients according to a defined set of therapy options,
- quantifies therapy change dynamics which often prove to be major drivers of market performance, and
- forecasts individual drug revenue specific to an indication.

To include individual treatment options in the Standard Template, a defined set of currently available and pipeline therapies must be established. For the sake of simplicity, this process will be demonstrated with an example that is simpler than most real-world pharmaceutical markets, but is easily extended or modified as needed.

Example Treatment Option Set

In a particular chronic disease market, there are three currently available therapies (A, B, and C) and one in-line product (D). Treatment A is used strictly as a monotherapy while Treatment C can be used as a single treatment or in combination with Treatment B. Treatment D will likely launch the market in early 2013 and will be used as a stand-alone therapy.

In defining the list of treatment options, simply tallying the current and future drug offerings does not give the full picture, as it fails to capture the combination use of Treatment BC. In reality, combinations must be analyzed as if they were stand-alone treatments, as together they comprise a physician's intended treatment for the patient.* At this point, the concept of a Mutually Exclusive and Collectively Exhaustive (MECE) competitive set should be established (4). MECE means that the treatment option set completely covers the spectrum of possible therapy choices, but does not double count patients who may be on more than one medication at a time. In a truly MECE treatment option set, a patient can be on one, and only one, defined therapy regimen at any particular time point.

Defining the MECE monotherapies is relatively easy; Treatment A and Treatment C exist currently and in the future, Treatment D will be a choice as well. Combinations such as Treatment BC can be listed as if they were individual treatment options. For completeness, a category labeled "Treatment Other" is

* In many markets, fixed dose combinations which combine two chemical agents in a single pill are increasingly popular, reflecting the fact that combination therapy can be viewed as a unique treatment option.

Table 7 Sample MECE Treatment Option Set

Monotherapy existing	Monotherapy pipeline	Combination existing	Catchall
A C	D	BC	Other

often included as a catchall for monotherapies or combinations that constitute a small portion of the market and are not categorized under other therapy regimen designations. In this case, the MECE treatment option set is expressed in Table 7.

MECE designations are important for a number of reasons, as they

- provide a framework for consistent data collection and analysis,
- avoid double counting of patients, a problem that often plagues forecasting efforts,
- reflect actual practice by treating combinations as individual therapy options,
- allow the dynamics of therapy changes to be categorized and quantified,
- incorporate data collected on the patient level, and
- support the projection of patient-level data to overall epidemiology data.

The MECE framework provides the most complete and flexible way to define the competitive set in a given market, at least for the purposes of Dynamic Modeling. Clearly, establishing an MECE set in this simple example was not too difficult. Complications arise when there is a high degree of combination usage, when there is a large number of me-too drugs, or when various drug classes are viewed much differently in the marketplace. These complications can be addressed with various adaptations of the Standard Template, however, and the basic MECE framework of categorizing treatment options represents "best practice" when implementing the Dynamic Modeling approach.

The stock/flow model structure can now be disaggregated to accommodate this newly established MECE set. First, the stock of Currently Treated Patients is broken down into segments, creating "slots" for each element in the treatment option set. Diagrammatically, the stock takes on a 3-dimensional appearance indicating the presence of treatment segments as shown in Figure 9.

The number of elements or "stacks" is not limited by software technology, but rather by the ability to collect and analyze data in a model that is both useful and usable. In our experience, 15 treatment options in the MECE set represent the practical upper limit for the number of therapy choices.* Some models have

* With the improved depth and breadth of patient databases in recent years (see chap. 6), it is now possible to obtain large enough samples to justify an increased number of therapies for inclusion in the MECE set. The tradeoff with this additional detail complexity, however, is a decrease in usability of the model. Care must be taken to ensure that the resulting simulation is not overly complex, hence restricting its use.

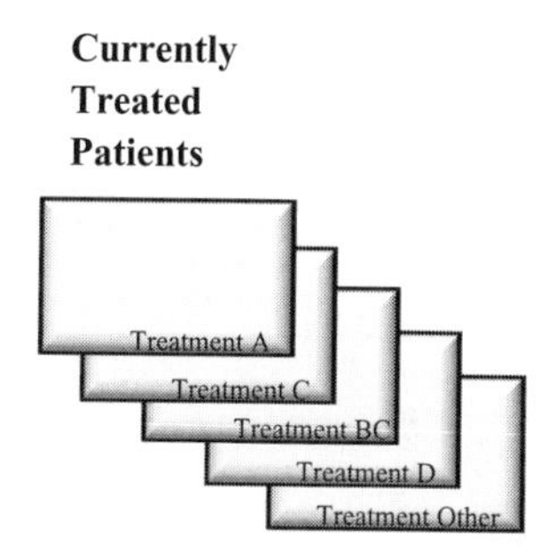

Figure 9 Disaggregating currently treated patients into MECE treatment set.

more, but as later chapters will demonstrate when addressing data collection and calibration options, adding treatment array elements exponentially increases the complexity of the modeling and data collection tasks.

With the MECE set established, the important dynamic of therapy changes can be included. Operationally, changes in treatment represent opportunities to capture patients and direct them to a specific therapy option. Recall that the stock of Currently Treated Patients is the basis for generating revenues in the Standard Template. A basic tenet of the Dynamic Modeling methodology, however, is that the number and distribution of Currently Treated Patients change as the result of the relative magnitude of the flows attached to it. Clearly, categorizing and quantifying these therapy changes is of vital importance from both a forecasting and strategic planning perspective. Establishing an MECE set as shown in Figure 9 allows the Standard Template to capture the operational dynamics which ultimately drive changing patient numbers.

KEY TREATMENT CHANGE DYNAMICS CAPTURED BY MECE FORMULATION

Treatment change dynamics include the following:

1. Switches: A to C and C to D
2. Add-ons: C to BC
3. Downgrades: BC to C
4. Titrations*: Increasing the dose of a drug but not adding or switching therapies

The Standard Template including treatment changes is shown in Figure 10.

* Titrations represent a market opportunity as indicators of dissatisfaction with current therapy, but do not move patients between therapy options within the Currently Treated Patients stock. As such, titrations will *not* be included as we extend the example Standard Template framework, but are easily incorporated using this methodology.

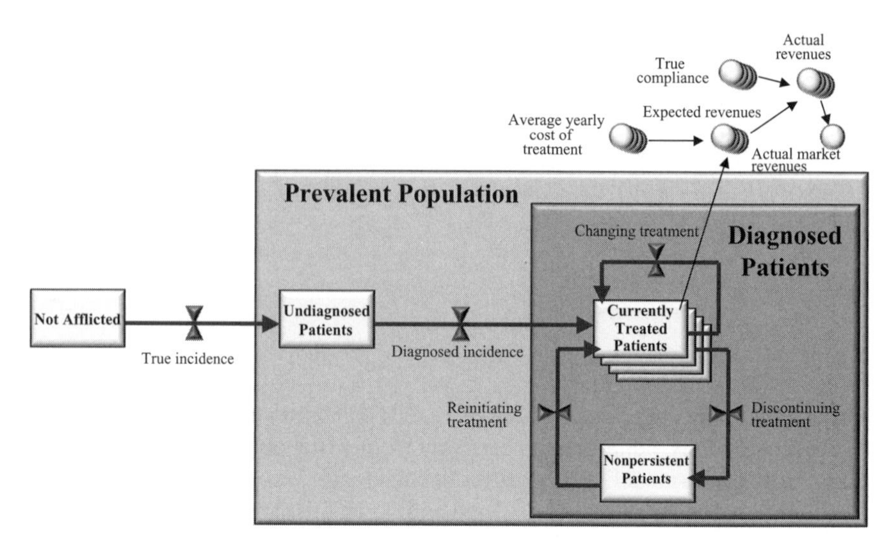

Figure 10 Treatment option revenue calculation structure using the Standard Template.

Some points to notice in this expanded diagram include the following:

- The stacked stock labeled Currently Treated Patients shows that patients are being tracked according to individual treatment options.
- The curious flow labeled *changing treatment* allows patients to move between elements in the MECE treatment set. In other words, these patients remain Currently Treated, but are transitioning from one therapy option to another.
- Each of the auxiliary variables can now be segmented by the treatments included in the MECE set. Treatment-specific data such as *average yearly cost of therapy* and *true compliance* can be used to calculate *actual revenues* for the individual drugs, which can then be summed up to arrive at *actual market revenues*. This process allows the Standard Template simulation to translate patients into drug-specific revenue projections. Clearly, combination usage must be taken into account when calculating revenues for a particular drug, such that the revenues for Treatment C include those generated by monotherapy patients as well as those on a Treatment BC combination.

The *changing treatment* flow occupies a small portion of Figure 10, yet captures some very important dynamics which ultimately drive market evolution. Patients *changing treatment* can go FROM any treatment option in the MECE set TO any other option over a given time period. As we extend the example Standard Template, assume that a primary market research project in 2009 determined the number of patients experiencing therapy changes throughout the course of the year.

Table 8 Example Treatment Change Matrix (2010)

FROM/TO	A	C	BC	D	Other	SUM
A	NA	15	23	0	2	40
C	22	NA	2	0	2	26
BC	5	5	NA	0	0	10
D	0	0	0	NA	0	0
Other	3	0	1	0	NA	4
SUM	30	20	26	0	4	80

These treatment dynamics can be conceptualized in a 2-dimensional Treatment Change Matrix, as shown in Table 8.*

Strategic Implications of MECE Treatment Option Sets

Developing a comprehensive group of therapy elements for a given disease indication allows a Brand Panning team to operationally identify the key treatment change dynamics which drive marketplace evolution. From a forecasting perspective, these patient flows result in the expected sales trajectory a newly released compound is projected to experience. From a strategic perspective, MECE treatment option sets define the patient flows which marketing strategies are designed to leverage, the results of which can be quantified by simulating their impact on the overall marketplace.

Using MECE Treatment Option Sets to Define Patient Flow Behavior

The rows of the Treatment Change Matrix represent the therapies switched FROM during 2009. For example, Table 8 indicates that 40 patients switched from Treatment A over the course of the year. The columns represent the therapies switched TO over the same interval. For example, the data in Table 8 shows that 30 patients switched to Treatment A over the course of 2009. Because our extended example does not include titrations, it is not possible for patients to switch from one drug to the same drug (cells labeled NA), but is an easy extension. Also note the lack of therapy changes associated with treatment D, as it is a pipeline drug and not yet on the market in 2009.

Treatment Change Matrices are concise visual representations of therapy dynamics driving market behavior, and careful examination of the data they contain can provide some interesting insights into marketplace dynamics.†

* The data in a Treatment Change Matrix can be expressed in percentage terms, which can then be input into a first-order Markov model. See chapter 10.

† Such switching behavior has become an important part of the analytics used to evaluate a number of nondurable consumer goods and services markets, including automobiles, cellular phones, and satellite radio/television.

KEY INSIGHTS FROM TREATMENT CHANGE MATRIX DATA

Careful analysis of change matrices can lead to a number of important insights, including

- comparison of the SUM categories gives an indication of net gain/loss of patients through the *changing treatment* flow. For example, 10 patients switched from Treatment BC but another 26 switched to it. This net gain of 16 patients, all else being equal, will translate into increasing patient share for Treatment BC over time;
- a therapy option can have no net gain or loss of patients, yet experience a significant amount of patient turnover. For example, "Treatment Other" experiences no net acquisition or loss of patients (gaining 4 and losing 4 each year), but has relatively high rates of therapy changes associated with it;
- analysis of intertreatment flows may indicate possible therapy progression. For example, 22 of the 26 switches from Treatment C are to Treatment A, suggesting a clear choice of subsequent therapy which replaces Treatment C;
- intertreatment dynamics may not be strictly linear or represent true lines of therapy. Although Treatment A generally follows Treatment C in the therapy progression, a significant number of patients (15 of them) switch treatments in the opposite direction; and
- magnitudes of switching dynamics may indicate relative drug utility as patients progress from one treatment option to the next. Patients switching from Treatment BC, for example, are evenly split as they go to Treatment A (10 of them) or Treatment C (10 of them). The concept of tying patient dynamics to relative therapy attractiveness will be addressed in later chapters, but this example should suggest these treatments are viewed in a very similar fashion for patients coming off combination therapy.

The details of how a Treatment Change Matrix can be populated with data, incorporated into the Standard Template, integrated with aspects of Doctor Adoption and Treatment Attractiveness, and analyzed with an eye toward strategy development will be addressed in later chapters. The Patient Flow dynamics regarding therapy changes are often in flux due to changing marketplace conditions, and the Dynamic Modeling framework is flexible enough to account for such changes. This explicit categorization of therapy dynamics provides a fuller explanation of the part of the Standard Template focusing on patients *changing treatments*, which in many chronic indications is the primary driver of patient shares and marketplace behavior.

The Standard Template at this point embodies some key epidemiology and market metrics of interest in many disease marketplaces. The implications of this framework and how dynamic model simulations can drive strategic insights will be explored as the methodology is extended and further detailed in the fully integrated approach. Before incorporating other aspects of the Standard Template, however, it will be useful to investigate how the dynamics of patient flow are of importance in understanding overall market behavior.

PATIENT FLOW ALLOCATION

The careful reader will have noticed that the diagram of the Standard Template now needs to be modified to reflect the new disaggregated structure involving the MECE treatment option set. In order for the Currently Treated Patients stock to be segmented, the model needs to allow for individual flows into and out of each treatment option. In other words, any inflow or outflow to the Currently Treated patients stock *must* be disaggregated in exactly the same dimensions; otherwise, there is no "pathway" to allocate patients into the correct treatment slot. Doing so changes the picture only slightly but has much larger implications, as shown in Figure 11.

First, the structure in Figure 11 now depicts the fact multiple flows into and out of each of segmented stock. In our continuing example, this means five flows—one for each current and future treatment option. Second, it should be apparent that the dynamic model can get very complicated very quickly, especially in the *changing treatment* flow. As was shown in the treatment change matrix, this flow needs to quantify both where patients come *from* and where they go *to*—in effect creating an 5 × 5 grid of potential dynamic pathways. Third, from a modeling perspective, we must now determine how to allocate the aggregate flows of patients across the various treatment options. The *diagnosed incidence* flow of *16*, for example, must now be distributed between our MECE treatment set. Similar allocations must take place for the *discontinuing treatment*, *reinitiating treatment*, and *changing treatment* flows. In general, there are three ways the Standard Template can allocate these patient flows.

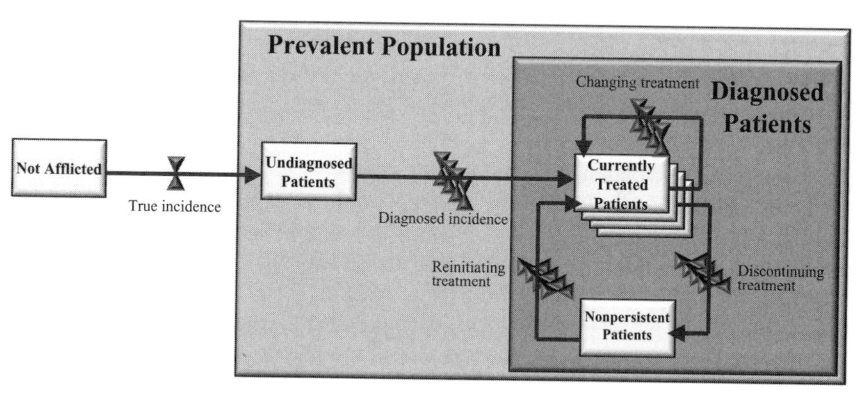

Figure 11 Structural representation of disaggregated flows in the Standard Template.

Table 9 Key Patient Parameters in Sample Dynamic Model (2010)

Treatment	Currently Treated Patients (*initial value*)	diagnosed incidence	changing treatment (*out*)	changing treatment (*in*)	reinitiating treatment	discontinuing treatment
Total	200	16	80	80	25	20
A	50	10	40	30	12	5
C	80	4	26	20	5	4
BC	60	0	10	26	3	6
Other	10	2	4	4	5	5

Current Data Applied to All Future Time Periods

One use of such a dynamic model is to investigate the implications on future patient distributions and/or drug-specific revenue generation assuming constant dynamics over time. In other words, the best available data can be collected on the aggregate flows of patients and how they are currently being distributed between the various treatment options in the MECE set. The model can then be simulated forward in time to show dynamic changes in the marketplace resulting from these parameters. Doing so can begin to give insight as to market direction based on currently available data.

Chapter 6 will detail recommended data collection methods for populating such a model, but for now assume that a comprehensive analysis of all available datasets gives the initial treatment distributions and patient flow allocations as shown in Table 9.

Note that even without a simulation model, the data in Table 9 begin to provide key strategic insights into this example marketplace.

1. Note the TOTAL *changing treatment (out)* is the same as the TOTAL *changing treatment (in)*. This must be true by definition, as in a defined MECE competitive set, patients are simply changing treatment options, not exiting treatment.
2. Note that more patients are *reinitiating treatment* (25) than are *discontinuing treatment* (20). All else being equal, this implies the number of nonpersistent patients is going down over time. Conversely, the Currently Treated stock will going up over time, as more patients are coming back to treatment than are discontinuing.
3. Careful analysis of the allocation of *diagnosed incidence* and the *changing treatment* flows can often show patterns of therapy progression. For example, no first-time patients start on Treatment BC, but many ultimately change to it later through the *changing treatment (in)* flow. Similarly, Treatment A is often used as first-line therapy, both for newly diagnosed patients and patients

returning to treatment, but the therapy change dynamics shows more patients switch from it (40 patients) than switch to it (30 patients).
4. Investigating the *discontinuing treatment* flows can elicit some interesting information about persistency. Why, for example, do 5 out of the 50 patients (10%) on Treatment A completely drop off treatment in a year, while only 4 out of 80 Treatment C patients (5%) do so?

To hand calculate the effect of these dynamics on the Currently Treated Patients stock over time is not entirely easy, and numerous studies have demonstrated the inability of humans to mentally determine the outcomes of even extremely simple dynamic systems.

To test your proficiency in mentally simulating dynamic behaviors, make a prediction as to the number of Currently Treated Patients on each of these five treatment options after six years have passed. For the sake of simplicity, assume the data shown in Table 9 are the *only* relevant aspects of the model—diagnosed prevalence is not changing, the flow metrics remain constant over time, and deaths are not included. Most people find the associated calculations become overwhelming very quickly, and simply guessing the ultimate behavior of the system is a shot in the dark (5). The relevant point of this exercise is worth emphasizing here: *The case and associated data shown in Table 9 is far simpler than* ***any*** *real-world pharmaceutical market, yet the effect of the example dynamics is extremely difficult to intuit or calculate correctly.* Yet, it is precisely the interaction of dynamics which operationally define how markets evolve over time, and correctly understanding those dynamics is vital for effective strategy development and/or forecasting accuracy. Fortunately, dynamic simulation tools are explicitly designed to capture and quantify these types of interactions and make the process of evaluating such dynamics a relatively simple task.

The advantage of Dynamic Modeling is shown by creating a stock/flow structure and inputting the parameters shown in Table 9. This structure can be built in a matter of hours, providing a quick and cost-effective way to evaluate resulting market behavior. Simulating the dynamic model through 2015 produces the results shown in Table 10.

In the example outputs shown in Table 10, Treatment C starts off as the market leader, but the dynamics of the system show it will soon be replaced by Treatment BC. Treatment A will soon pull past it as well. These results are not easy to intuit simply from the static dataset, and the Standard Template simulation quickly and easily shows outputs which can be discussed and analyzed.

The reader should not conclude that such results are "right" or even realistic—the future is unlikely to behave *exactly* like the past. In addition, this example analysis fails to take into account how the dynamics of the system will change with the entry of in-line products (Treatment D). However, even this simple case shows how the power of simulation technology, coupled with real-world data, can add insight to both current market environment and possible evolution.

Table 10 Patients on Individual Treatment Outputs from Dynamic Model (End of Calendar Years 2009–2015)

Treatment	2009	2010	2011	2012	2013	2014	2015
Total	200	221	242	263	284	305	326
A	50	57	64	71	77	85	92
C	80	79	78	77	76	75	74
D	0	0	0	0	0	0	0
BC	60	73	86	99	112	125	138
Other	10	12	14	16	18	20	22
Total (%)	100	100	100	100	100	100	100
A (%)	25	26	26	27	27	28	28
C (%)	40	36	32	29	27	25	23
D (%)	0	0	0	0	0	0	0
BC (%)	30	33	36	38	39	41	42
Other (%)	5	5	6	6	6	7	7

End-user Input Scenario

To address some of the drawbacks of approach No. 1, the Standard Template can instead be configured to allow users to define what the allocation of various inflows and outflows will be over time. In doing so, the model continues to show implications of patient-movement dynamics, but these parameters now become scenario variables rather than static data points which stay constant over time. While these scenario variables are not data-based per se (there is no "data" about the future), including them helps provide insight regarding market evolution and makes the Standard Template a much more effective scenario planning tool and strategy development platform.

In making this transition, the absolute magnitudes of individual flows are often changed into percentage terms. The dynamic model can be formatted to allow these percentages to be changed over time, while ensuring, it does not "create" patients.) To do so, the numbers from Table 9 can be converted into component percentages as shown in Table 11.

Table 11 Percent Allocation of Dynamic Model Flows (2010)

Treatment	Currently Treated Patients	diagnosed incidence	treatment changes (out)	treatment changes (in)	reinitiating treatment	discontinuing treatment
Total	200	16	80	80	25	20
A	50	62.5%	50.0%	37.5%	48.0%	25.0%
C	80	25.0%	32.5%	25.0%	20.0%	20.0%
BC	60	0.0%	12.5%	32.5%	12.0%	30.0%
Other	10	12.5%	5.0%	5.0%	20.0%	25.0%
SUM (%)		100	100	100	100	100

Table 12 Sample Distribution of *diagnosed Incidence* in Standard Template Model

Treatment	2010	2011	2012	2013	2014	2015
A (%)	62.5	62.5	60	58	55	51
C (%)	25	25	27	26	28	30
D (%)	0	0	0	5	7	10
BC (%)	0	0	0	0	0	0
Other (%)	12.5	12.5	12.5	11	10	9
SUM (%)	100	100	100	100	100	100

In other words, the dynamic model can be established such that the *diagnosed incidence* flow, for example, is always 16, but the distribution or allocation of that flow across the various treatment options can vary over time, as shown in Table 12.

The model end-user in this case can choose to have the distributions stay constant over time or can change the allocations to test the impact of potential marketing strategies. For example, what would the resulting patient numbers be if Treatment D could capture 5% of newly diagnosed patients in its first year of launch (2013), 7% in 2014, and 10% in 2015? Similar scenarios can be run against the other flows in the model such as *discontinuing treatment, reinitiating treatment, changing treatment (out)*, and *changing treatment (in)* to test their impact on the patient numbers for any of the defined treatment options in the Standard Template.

Such an analysis can prove extremely valuable by:

- *Providing a simple way to evaluate basic sensitivity of potential market strategies.* "Should we go after newly diagnosed patients or patients returning to treatment?"
- *Delivering insight as to feasibility of strategic forecasts.* "Even if we capture 40% of all patient movements in our first two years of launch, we will only achieve a 25% patient share within 24 months."
- *Allowing for basic strategies can be combined and evaluated.* "What if we captured 50% of all patients changing therapy *and* were 20% better at keeping our patients from becoming Non-Persistent?"

The addition of end-user input scenarios into the Standard Template results in a powerful tool for market evaluation and what-if analysis. Many of our clients have used models no simpler than this, combined with consensus and reasonability assumptions about the share of patient flows a product might capture, to help develop more realistic strategic plans and corresponding targets for their pharmaceutical products.

Allocation Algorithm

Extending a basic input-based model to one where treatment allocation across flows is the result of a mathematical choice algorithm is a path a number of clients have chosen to pursue. Chapter 10 deals with such algorithms in detail. For now, the reader should be aware that the choice algorithms draw on published research; integrate concepts of treatment attractiveness; and doctor adoption and are calibrated to specific market-level data. Adapting the Standard Template to include such advanced analysis adds a level of rigor and sophistication which is often beneficial to client teams. The inclusion of flow allocation algorithms is a nice but not necessary component of the Dynamic Modeling approach, and many teams have used models no more complex than those described in this chapter to gain valuable market insight. The sophistication of the flow allocation methodology depends on the questions the dynamic model is designed to address, and hence varies by company, time frame, and even indication.

Strategic Implications of Patient Flow Allocation

Dynamic models give the members of a Brand Planning team a tool to test various scenarios and possible strategies in a risk-free simulation environment, based on operationalizing the patient dynamics driving the evolution of an indication marketplace. Rigorous testing of how flows of patients might be allocated between various treatment options in the MECE set gives insight into how the market might change over time, how various marketing strategies might impact those changes, and what patient flow allocations leverage the marketplace dynamics effectively. Such "what if" analysis provides a means to quantify the impact of a potential strategic decision before it is being implemented in the real world.

The Standard Template dynamic model provides a useful methodology to begin analyzing the combination of epidemiology and treatment dynamic factors which drive market behavior and subsequent revenue generation in a given indication. We will return to the use of this framework for strategy development in chapter 11, using the continuing example structure established in this chapter and extended in later ones. Readers should be reminded that Dynamic Modeling is an approach rather than a set of fixed equations which are "parameterized" for a given marketplace. The Standard Template has proven to be an effective and extendable basic framework in many chronic indications, but does not represent the only means through which the Dynamic Modeling approach can be used to analyze a disease indication.

SUMMARY

- Standard Template framework provides a structured methodology to address chronic indications with defined treatment options and high degrees of patient churn.
- Stock/flow structure explicitly captures epidemiology concepts of incidence, prevalence, and mortality.
- Dynamic model establishes strict definitions of compliance and persistence which uniquely represent patient characteristics and their revenue impact.

- Treated patients can be divided between a MECE treatment option set, allowing patients to be translated into revenues.
- MECE set allows for categorization and quantification of key intertreatment dynamics.
- Treatment Change Matrices provide a useful visual representation of the dynamics involved as patients switch treatment.
- Stock/flow structure captures other important treatment dynamics of discontinuation and reinitiation.
- Simulation of Standard Template model allows for testing of patient flow allocations to determine dynamic behavior within an indication.

REFERENCES

1. Improving Chronic Illness Care—A National Project of the Robert Wood Johnson Foundation. http://www.improvingchroniccare.org
2. Chronic Illness Coalition. http://www.home.tir.com/~slater/id1.html.
3. Centers for Disease Control and Prevention: National Diabetes Fact Sheet: general information and national estimates on diabetes in the United States, 2003. Revised edition. Atlanta, GA. Centers for Disease Control and Prevention, U.S. Department of Health and Human Services.
4. Rasiel EM. The McKinsey Way: Using the Techniques of the World's Top Strategic Consultants to Help You and Your Business. New York, NJ. McGraw-Hill, 1999.
5. Sweeney LB, Sterman JD. Bathtub Dynamics: Initial Results of a Systems Thinking Inventory. Working paper, v. 1.2. September, 2000. Forthcoming, System Dynamics Review. http://www.web.mit.edu/jsterman/www/bathtub.
6. Boyle JP, Honeycutt AA, Narayan KM, et al. Projection of diabetes burden through 2050: Impact of changing demography and disease prevalence in the U.S. Diabetes Care 2001; 24(11):1936–1940.
7. Bernstein PL. Against The Gods: The Remarkable Story of Risk. John Wiley & Sons, 1998:74–87.

5

Extensions of the Standard Template

INTRODUCTION

This chapter will cover the following topics:

- Expansion of diagnosis process
- Inclusion of nonchronic diseases
- Categorization of misdiagnosis
- Progression of lines of therapy
- Disaggregation of model structure into more detailed patient groups

In our consulting practice, the Standard Template is often used to introduce clients to the Dynamic Modeling methodology and how it can be applied to chronic indications. However, the flexibility of the approach allows this basic structure to be adapted and expanded to meet the unique characteristics of individual markets. This chapter will focus on five common modifications to the Standard Template, which have proven themselves useful in analyzing various indications.

STAGES OF DIAGNOSIS

In some chronic indications, initial epidemiology data show a large Undiagnosed Patient population. From a stock/flow perspective, such a situation indicates that these patients spend a long time, on average, before being correctly diagnosed. In the case of a 30% diagnosis rate, for example, the basic epidemiology could be populated as shown in Figure 1.

Clearly, the Standard Template can be applied to the Diagnosed Patients stock in the manner previously described in chapter 4. But the Undiagnosed Patients stock in this case stands out as an area begging for further explanation. For this 70/30 distribution to exist, there must be some "bottleneck" or series of obstacles preventing patients from being correctly diagnosed.* Expanding the

* Debottlenecking is a common technique applied to remove capacity constraints in production processes, and can easily be adapted to patient flow bottlenecks in diagnosis processes or health care delivery settings such as hospitals or emergency rooms.

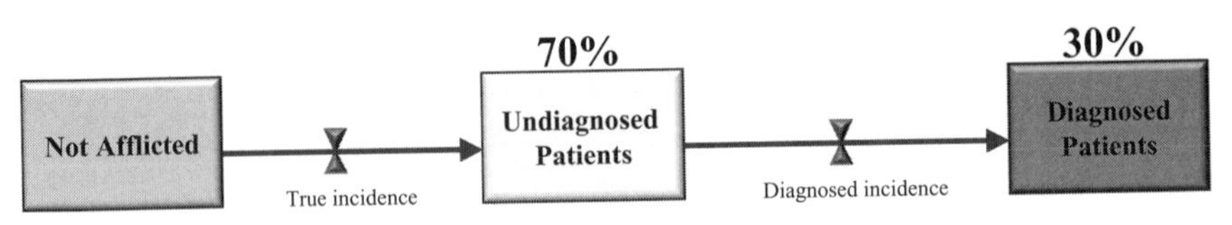

Figure 1 Stock/flow epidemiology framework depicting 30% diagnosis rate.

Standard Template structure to examine the stages or steps a patient must go through to receive a correct diagnosis may be of strategic importance.

Case Study: Expanding Undiagnosed Patients Stock

Imagine a disease that primarily afflicts adolescents and is difficult to diagnose. The symptoms of the condition are usually ignored or misinterpreted, resulting in a very low diagnosis rate. Once identified, the diagnostic process is long and not physiologically definitive (i.e., the disease is mostly behavioral, so there is no blood test or imaging technology that can readily identify it). Disorders such as schizophrenia, attention deficit and/or hyperactive disorder (AD/HD), and even adolescent depression generally exhibit these characteristics, and would thus be candidates for the dynamic market structure that follows.

To investigate the Undiagnosed Patients stock in more detail, it is often useful to identify the phases or stages a patient goes through before being correctly diagnosed. The example structure shown in Figure 2 is by no means definitive, but has proven useful with a number of our clients in analyzing diagnosis processes.

As depicted in Figure 2, patients in this example indication first develop the disease physiologically before even being aware that there is a problem, necessitating an Unrecognized Symptoms stage. Eventually, someone becomes aware of a behavioral problem. This example case includes structure indicating that the patient himself, his parents, or a school worker/friend could be the first to recognize some symptoms, without necessarily knowing the disease associated

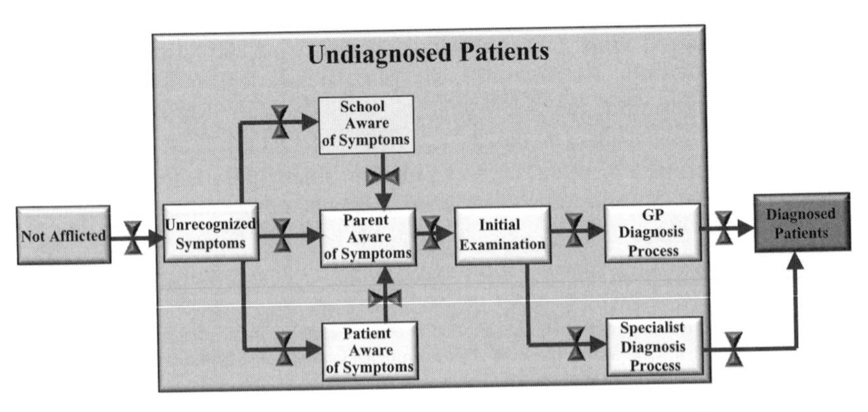

Figure 2 Stock/flow framework of undiagnosed patient structure expansion.

Table 1 Example Average Time Spent in Various Diagnosis Stages

Stage	Average time in stage (months)
Unrecognized Symptoms	18
Patient Aware of Symptoms	6
Parent Aware of Symptoms	3
School Aware of Symptoms	2
Initial Examination	1
GP Diagnosis Process	12
Specialist Diagnosis Process	6

with them. Since the parents ultimately decide and schedule a visit to a doctor, eventually they must be made aware of the patient's symptoms, even if notified by the sufferer or someone from his school. An initial Examination will eventually be scheduled when the parents become aware of the problem, possibly after some delay. The patient may be referred to a specialist or the Diagnosis Process may be headed by a general practitioner (GP). Eventually, however, the patient receives a correct diagnosis for the affliction and enters the stock of Diagnosed Patients. For simplicity, mortality flows have not been shown in Figure 2 but are usually incorporated for completeness.

The detail of this Dynamic Modeling structure allows the various stages and times associated with the stock of Undiagnosed Patients to be identified and quantified. In aggregate, Figure 2 represents the possible stages and dynamic pathways patients must pass through between disease initiation (*true incidence*) and disease definition (*diagnosed incidence*).

The data needed to populate such a dynamic structure is relatively straightforward. First, the average time spent in each stage or phase needs to be determined. These data can be collected from published data sources, secondary market analysis, or even primary market research to fill in the gaps. Data on stages and their associated times can be collected and organized as shown in Table 1.*

Second, for stocks with multiple outflows, the percentage of patients taking each of a defined number of possible exit paths is required. For example, what percentage of patients go through the GP Diagnosis Process versus a Specialist Diagnosis Process? These types of estimates are more difficult to quantify but are often obtainable through detailed patient records. For example, a questionnaire completed by a treated patient regarding the onset of symptoms and their recognition might reveal the data shown in Table 2.

* Sophisticated modeling concepts such as discrete versus ordered delays can be combined with average time spent in each defined phase of the diagnosis process, if the need arises. (see Sterman's *Business Dynamics*).

Table 2 Example Distribution of Exit Paths from Unrecognized Symptoms Stage

TO:	FROM: Unrecognized symptoms
Patient Aware of Symptoms	10%
Parent Aware of Symptoms	20%
School Aware of Symptoms	70%

Similarly, a survey of physicians, particularly specialists in the disease area, should reveal the number of patients initially diagnosed by a GP instead of a specialist, as shown in Table 3.

The visual nature of stock/flow representations is especially valuable in these types of cases, as the diagram easily accommodates various input data, as shown in Figure 3.

Even without simulating the dynamic model shown in Figure 3, the strategic implications of this structure should be readily apparent:

- If the goal is to decrease the time consumed in the entire process, then the stages in which patients spend the greatest amount of time represent the bottlenecks to the system. The values of 18 months before symptom recognition and 12 months for the GP Diagnosis Process leap out here.
- Note how the distributions of the outflows affect the system. 70% of patients are "funneled" to a GP, where they spend a year, on average, before being diagnosed. Diagnosis time is cut in half if they are directed to a Specialist.
- A similar situation exists regarding awareness of symptoms, whereby if the Parent is the first to recognize something is wrong, less time is consumed before an Initial Examination is scheduled.

The outputs from dynamic flow chart models can be hugely valuable, and *they provide a level of market analysis, which is not possible without the use of simulation*. First, this model can be used to determine time metrics associated with Undiagnosed Patients, which are as follows:

Table 3 Example Distribution of Exit Paths from Initial Examination Stage

TO:	FROM: Initial examination
GP Diagnosis Process	70%
Specialist Diagnosis Process	30%

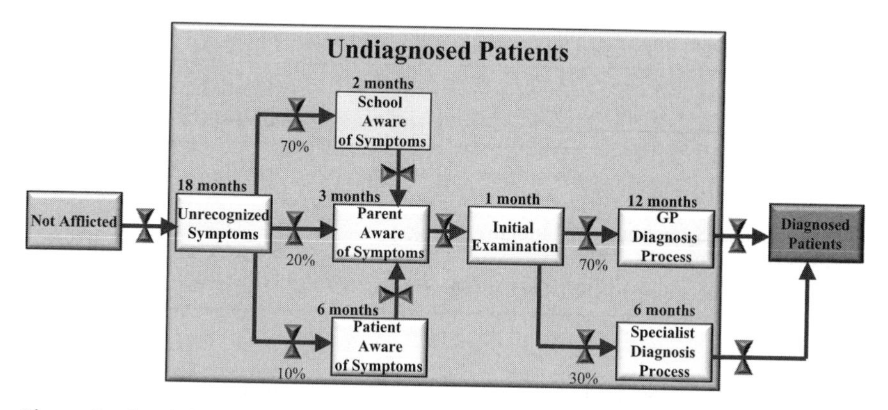

Figure 3 Stock/flow framework of undiagnosed patient structure expansion with associated example parameter data.

- How long, on average, does it take for patients to be diagnosed?
- How many patients complete the process in fewer than 12 months? Or, less than 24 months?
- How many patients die before they are *ever* diagnosed?

Second, the dynamic model can be used to calibrate the distribution of patients between the various stocks at any given point in time. This analysis is only possible with a Dynamic Modeling approach and provides insight not possible with static evaluations. This type of stage quantification allows marketing teams to identify the largest pools of potential patients and where they are in the diagnosis process. In this example case, the distribution of patients across the stages to diagnosis can be calculated using the dynamic model, the results of which are shown in Figure 4.

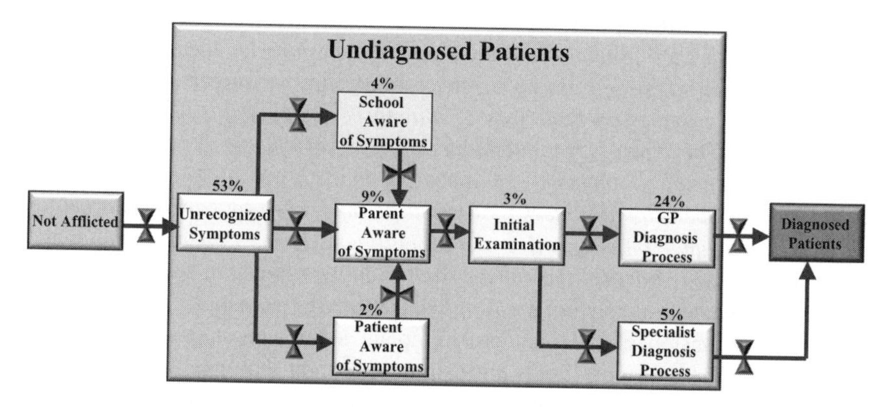

Figure 4 Expanded undiagnosed patient structure with derived stock distribution.

Table 4 Example Impact of Parameter Changes on Diagnosis Rate

	Resulting diagnosis rate
Base Case	30%
Time in:	
Unrecognized Symptoms = 14 months	32.7%
GP Diagnosis Process = 8 months	31.8%
Percent of patients going to GP Diagnosis Process = 30%	31.5%

Third, the dynamic model provides a tool to evaluate the parameters that govern this process. From a strategic standpoint, such a simulation allows some very interesting questions to be answered:

- What is the impact of changing the average time it takes for someone to recognize the symptoms of the disease, or for a GP to diagnose it?
- How would the behavior of the system change if more patients went to a Specialist for diagnosis?
- Would encouraging parents to schedule Initial Examinations more quickly and significantly change diagnosis rates?

Evaluating the sensitivity of the dynamic model to changes in the dynamic parameters, either in isolation or in combination, is a great starting point for strategy development. For example, Table 4 shows how diagnosis rates would change based on simple parameter changes.

Strategic Implications of Dynamic Modeling of Diagnosis Stages

Dynamic Modeling simulation results provide insight into high-leverage intervention points for changing diagnosis rates. The magnitude, feasibility, and tactics associated with such potential strategic leverage points can then be evaluated to determine an integrated market strategy. Analyzing the stages through which patients must progress on their way to diagnosis has an important conceptual result as well. This type of Dynamic Modeling exercise changes the diagnosis rate parameter in an indication from a set input *to a dynamic* output*—that is the result of a series of operational phases that might be influenced by various marketing actions. The change in perspective regarding diagnosis rates opens a whole new avenue of strategic possibilities for a Brand Planner, placing at least one aspect of disease epidemiology within the range of marketing influence.*

The expansion of the Standard Template to include stages of diagnosis has provided many of our clients with strategic insight that goes beyond a simple epidemiology evaluation or sensitivity analysis. The operational nature of the stock/flow structure ensures internal consistency of data inputs and outputs, resulting in a powerful simulation tool that can be used to test potential marketing strategies for Brand Plan development.

NONCHRONIC DISEASES

In chapter 4, the Standard Template was presented as dealing with chronic diseases from which there is no "exit" other than death. A patient with rheumatoid arthritis, for example, is always characterized as such. However, a simple extension of the Standard Template creates structure to address diseases from which recovery, and perhaps subsequent relapse, is possible. Afflictions such as cancer, depression, and pneumonia are all indications of this type.

First, since these types of diseases are not lifetime afflictions, the dynamic model must allow for a way to capture flows of patients recovering from the illness. One way to do so would be to simply flow them back from the stock of Prevalent Population to the stock of the not afflicted population. However, doing so would mask an important dynamic; many such maladies have higher incidence rates for recovered patients versus those who have never suffered from the illness before. For that reason, we find the structure shown in Figure 5 to be a good starting point for analysis.

Notice the addition of the Previously Had Disease stock in Figure 5 allows for patient flows of *recovery* and *relapse* to be included. This extension changes the prevalence categorizations, allowing the model to account for Lifetime Prevalence –(the number of people who Have Ever Had Disease) and Point Prevalence –(the

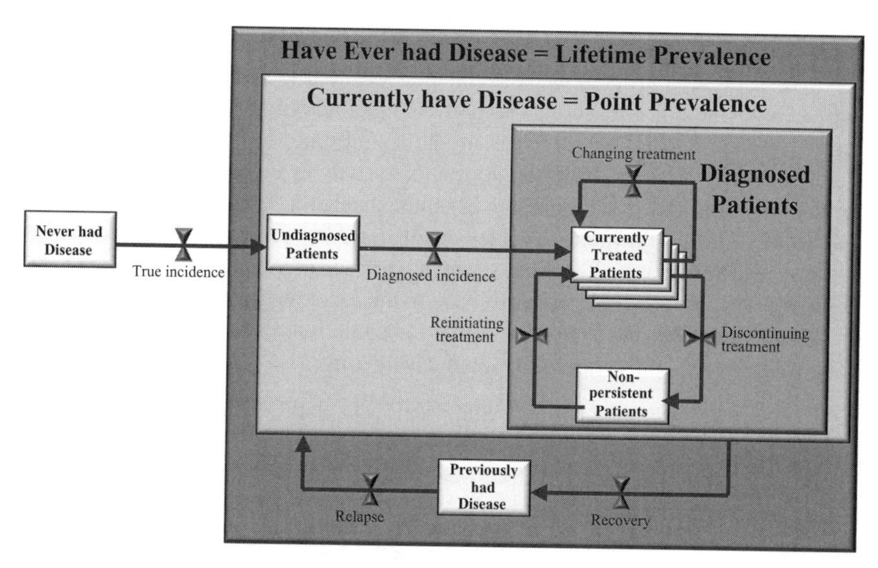

Figure 5 Expansion of Standard Template for nonchronic illnesses.

number of people who Currently Have Disease). In actuality, the flows of *recovery* and *relapse* come into or out of each of the stocks in the point prevalence category, but are shown in Figure 5 as aggregate flows for the sake of simplicity. These distinctions are usually supported by clinical data about relapse rates versus first-time occurrences of the disease. Additionally, this structure can still incorporate dynamics of therapy switches and treatment discontinuation/reinitiation in the Currently Treated Patients stock as was described in chapter 4.

The flexibility of this extended model can provide additional strategic insight into nonchronic diseases by explicitly showing and subsequently quantifying some key dynamics:

- Spontaneous recovery rates (patients being cured without drug intervention) can be compared to therapy-induced rates (patients coming from Currently Treated to Previously Had Disease), both in terms of absolute magnitude and probabilities.
- Metrics related to the average time the disease is in relapse before symptoms reoccur (and patients re-enter the Point Prevalence stock via the *relapse* flow) can be included as inputs or calculated as model outputs.
- Recovered patients can be tracked according to what previous treatments they received, so that recurrence could trigger past treatment affinity (patients return to the most recent and most effective treatment) or past treatment aversion (patients return to any treatment *other* than those previously received).

Our experience with client teams is that the parameters that govern the epidemiology of such nonchronic conditions are rarely known and often misunderstood. Integrating various estimates into a simple stock/flow model can provide some clarity about the nature of the disease and confidence in the metrics that ultimately drive key epidemiology flows.

To continue our example from chapter 4, suppose through data collection and the associated model outputs the structure defined in Figure 5 with produced results for a particular calendar year, as shown in Figure 6.

Note that most of the dynamic model outputs in Figure 6 are the same as those of our continuing example that began in chapter 4. However, the model now quantifies Lifetime Prevalence Values (600) and the stock of those patients who previously had disease (200). The associated flows of *recovery* and *relapse* are now populated as well. These quantified dynamics, coupled with outputs regarding the number of associated patients, allow for some interesting strategic insights, which are possible only through Dynamic Modeling:

- There are half as many patients who Previously Had Disease (200) as there are patients who Currently Have Disease (400).
- Patients spend, on average, four years in the Currently Have Disease stock before experiencing *recovery**.

* See chapter 3 for a review of stock/flow magnitudes versus average residency time.

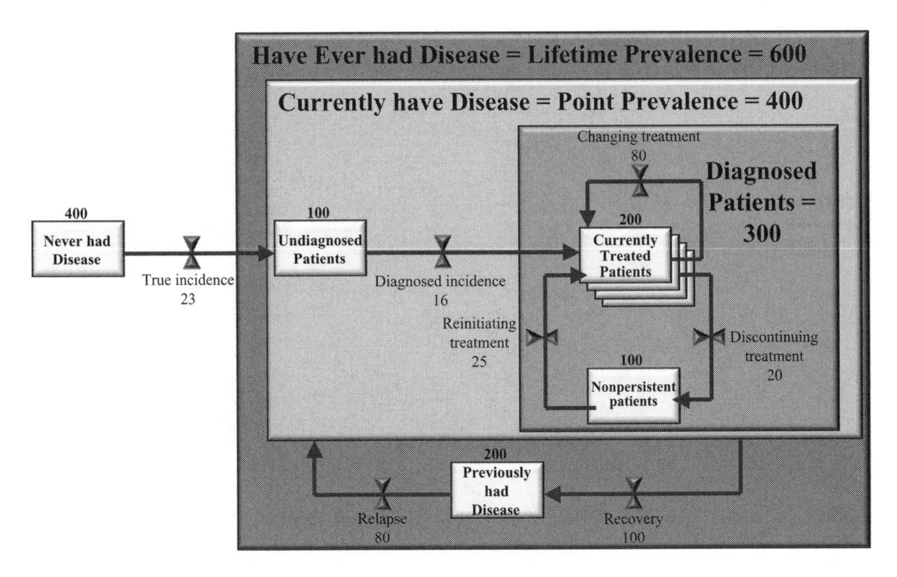

Figure 6 Expansion of Standard Template for nonchronic illnesses with associated sample data.

- Patients spend, on average, two and a half years in the Previously Had Disease before experiencing *relapse*.
- The flow of *relapse* patients (80) is significantly greater than the flow of *true incidence* (23).

Recall that the strategic decisions, which ultimately drive product sales, must focus on the *flows* of patients in the system, and that the stocks of patients will be the cumulative end result of those changing dynamics. From a strategic standpoint, the implications of the example shown in Figure 6 should suggest areas of focus for any new market entrant or existing product. This combination of epidemiology metrics and treatment dynamics creates a more vivid and illustrative picture of nonchronic indications, and subsequent analysis using the Dynamic Modeling approach often provides strategic insight which is not possible with traditional static market evaluations.

Strategic Implications of Dynamic Modeling of Nonchronic Diseases

Indications exhibiting recovery and subsequent relapse of patients are often difficult to analyze at the patient level due to their inherent dynamic nature. Yet categorizing and quantifying complex patient movement is precisely what dynamic models are designed to do. Furthermore, having an integrated platform in which such dynamics can be sanity-check provides an additional level of rigor. The Dynamic Modeling approach to nonchronic diseases ensures that patient epidemiology flows are internally consistent, providing a level of insight into disease

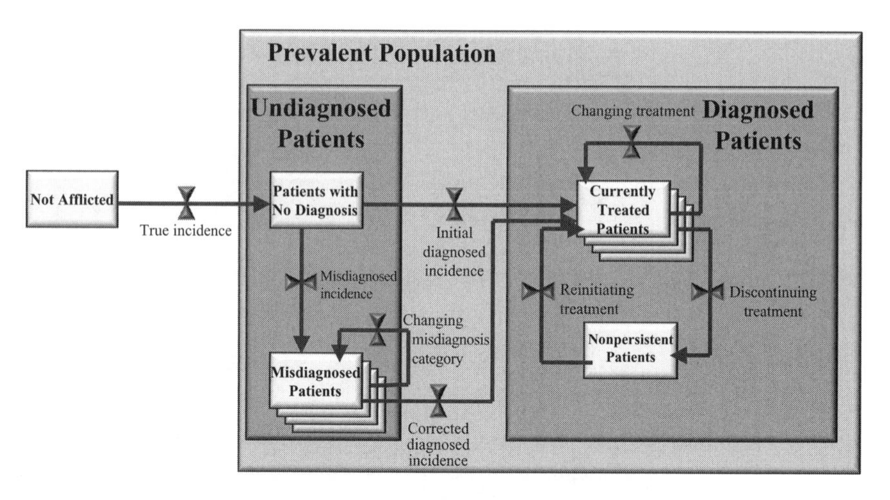

Figure 7 Expansion of Standard Template for inclusion of misdiagnosis.

dynamics that is vital for effective strategic planning. Combining recovery/relapse patient behavior with the treatment dynamics embedded in the Standard Template creates an integrated, operational approach for the analysis of nonchronic indications.

MISDIAGNOSIS STAGE

With some difficult-to-diagnose diseases, misdiagnosis of the true illness is a serious issue. Patients actually suffering from mononucleosis, for example, are often misdiagnosed as having depression or chronic fatigue syndrome. The basic stock/flow structure of the Standard Template is easily expanded to include the dynamics surrounding misdiagnosis, as shown in Figure 7.

Note that the structure shown in Figure 7 expands Undiagnosed Patients to detail those patients who have been Misdiagnosed and those Patients with No Diagnosis of the disease being modeled or of the misdiagnosis being set. In this structure, there are two flows into the stock of Diagnosed Patients, patients who were previously Misdiagnosed (*corrected diagnosed incidence*), and those who had no previous diagnosis (*initial diagnosis incidence*). Additionally, this new structure incorporates a *misdiagnosed incidence* flow; patients being mistakenly diagnosed with another disease rather than the one from which they actually suffer. The stock of Misdiagnosed Patients has been disaggregated to include a set of common misdiagnoses that allows the model to capture the dynamic of patients progressing through one or more misdiagnoses through the *changing misdiagnosis category* flow before eventually receiving a correct diagnosis.

By including the important dynamics associated with misdiagnosis, the Standard Template can be extended to better represent the epidemiology of diseases in which patients are often incorrectly categorized. A fully populated dynamic model with the structure detailed in Figure 7 can help answer some important strategic questions, such as the following ones:

- What is the number and distribution of patients across the key epidemiologic categories?
- How is the epidemiology expected to change over time?
- How many patients are diagnosed correctly at the first time (*initial diagnosed incidence*) versus misdiagnosed patients (*misdiagnosed incidence*)?
- How long, on average, do patients stay as Misdiagnosed Patients before being correctly diagnosed (*corrected diagnosed incidence*)?
- How are patients distributed between key categories of Misdiagnosis?
- Do patients progress through a series of Misdiagnosis categories before being correctly diagnosed?
- How many patients die before being properly diagnosed with their affliction?
- By how much would the flows of *initial diagnosed incidence* and *corrected diagnosed incidence* need to be changed to significantly alter the number and distribution of patients between Diagnosed, Misdiagnosed, and No Diagnosis

These key epidemiologic concepts can be integrated with the treatment dynamics structure of the Standard Template to have a more complete operational picture of the interactions driving the behavior of markets characterized by high levels of misdiagnosis.

Strategic Implications of Dynamic Modeling of Misdiagnosed Stage

Indications exhibiting high rates of misdiagnosis are often subject to epidemiology errors, as estimating how many patients are Misdiagnosed and the dynamics surrounding them is a difficult proposition. But from a strategic perspective, quantifying such metrics is vital for developing effective Brand Plans. Dynamic Modeling can determine the impact of high misdiagnosis rates on marketplace dynamics, as well as estimates the impact of changing said rates and the expected sales trajectories of future or existing indication offerings. This analysis can then be compared to the expected costs of such a misdiagnosis strategy, such as an education campaign to inform patients/physicians on indication symptoms, to determine the financial feasibility of such an effort.

The other category of misdiagnosis, involving patients included in prevalence estimates but who are not true sufferers of the disease, is also easily captured using the Dynamic Modeling approach. Figure 8 shows a stock/flow framework for conceptualizing both types of misdiagnosis—a structure that can easily be included in the extended Standard Template.

PROGRESSION OF LINES OF THERAPY

Chapter 4 detailed how treatment changes could be incorporated into the Standard Template with a simple Treatment Change Matrix that categorized patients switching between therapies. The example in that chapter works well in markets

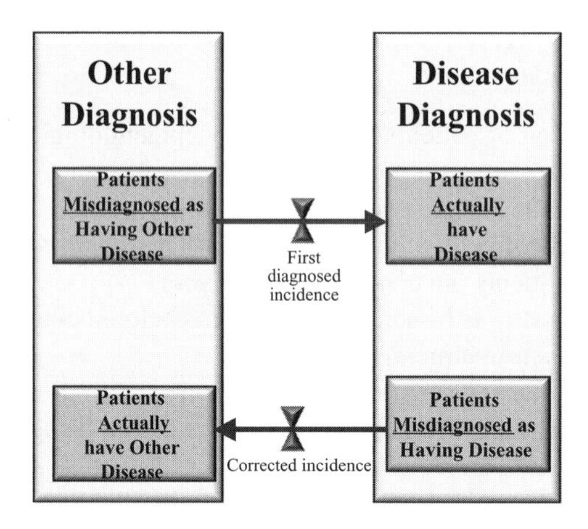

Figure 8 Stock/flow framework for categorizing two types of misdiagnosis.

where such treatment changes are fairly distributed; patients are as likely to switch from Treatment A to Treatment C, for example, as vice versa. In some markets, however, a basic FROM/TO formulation might mask some interesting patient dynamics.

Case Study: Lines of Therapy Dynamics

Imagine a market research team collected data on the number of patients switching between four therapy options in a given month, as shown in Table 5.

Data in the form of Table 5 can be extremely valuable in understanding the patient dynamics driving the behavior of a particular market. Some interesting aspects of the 1700 treatment changes each month should be readily apparent by examining this data.

- 70% of treatment switches *from* Treatment A were to Treatment C.
- 80% of treatment switches *from* Treatment C were to Treatment BC.
- 100% of switches *from* Treatment BC were to the treatment "other."
- There were 0 switches *to* Treatment A.
- Treatment Other is a salvage treatment, as evidenced by the fact that no patients switch *from* it to any other treatment.

Table 5 Example Patient Distribution and Treatment Change Matrix

Initial number of patients	FROM/TO	A	C	BC	Other	SUM
10,000	A	N/A	700	300	0	1000
4000	C	0	N/A	400	100	500
1000	BC	0	0	N/A	200	200
600	Other	0	0	0	N/A	0
15,600	SUM	0	700	700	300	1700

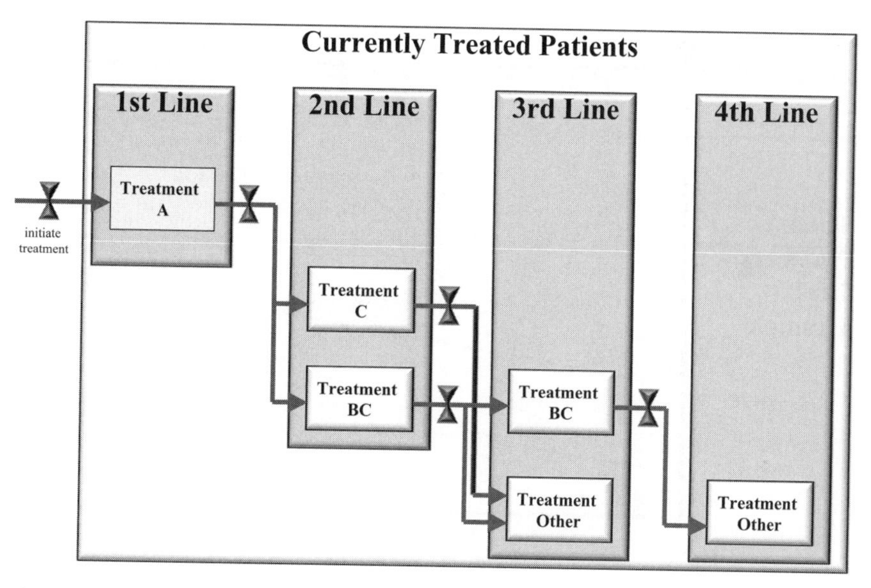

Figure 9 Expansion of Standard Template to capture lines of therapy progressions.

The switching data in Table 5 clearly shows a defined progression of treatments, in which patients generally go from Treatment A to Treatment C to Treatment BC to Treatment Other. And while these switching dynamics could certainly be incorporated in the basic treatment changes framework presented in chapter 4, a more detailed stock/flow picture that visually depicts this progression of therapy is both more accurate and illustrative. Such an extension will explicitly capture various **lines of therapy** or sequential therapy choices as patients move through a defined treatment progression. Figure 9 shows how the Currently Treated Patients stock of the Standard Template can be extended to incorporate such lines of therapy progressions* .

For simplicity, Figure 9 shows Treatment A as the only listed first-line therapy option, although this assumption should be supported by data analysis to determine the distribution of patients across various therapy regimens as they *initiate treatment*. A careful thought experiment using the data in Table 5 should reveal why Treatment A is at least one of the true first-line therapy options. The Treatment Change Matrix in Table 5 suggests that no patients switch *to* Treatment A, regardless of their current therapy. Yet there are 10,000 patients currently receiving that treatment option. These patients had to come from somewhere, and by using the operational nature of the stock/flow framework, it should be clear that they must have entered the Currently Treated Patients stock through the *initiate*

* For the sake of simplicity, the dynamics around deaths, treatment discontinuations, nonpersistence, and subsequent reinitiation of treatment are not shown in Figure 9 in chapter 5, but are easily incorporated as shown in chapter 4.

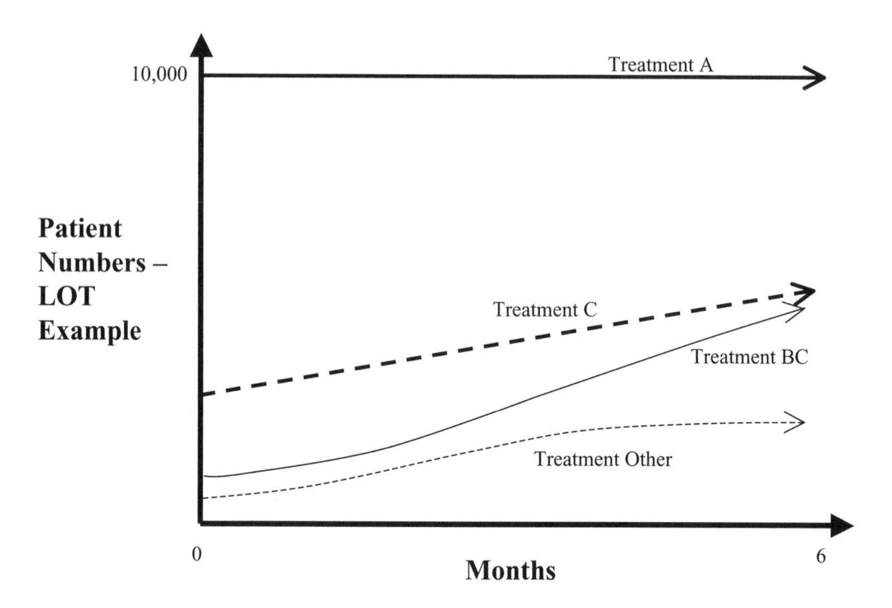

Figure 10 Example outputs from dynamic model capturing explicit lines of therapy progression.

treatment flow and entered the Treatment A stock directly*. Similar logic, applied left to right in the stock/flow structure, also explains why *Treatment Other* is not a listed as second-Line therapy option. In most real-world markets, lines of therapy are rarely as "clean" as in this simplified example. But the process of examining data and rendering it into the Dynamic Modeling methodology is one which often clarifies the murky picture of lines of therapy progression in many pharmaceutical marketplaces.

The explicit representation of these lines of therapy progressions allow for the calculation of the patient dynamics, which drive changes in the marketplace. It also provides key insights into changing patient numbers that might not be clear from simply looking at the associated switching dynamics. Data for the rate at which patients *initiate treatment*, combined with the appropriate switching parameters along lines of therapy, allow a dynamic model to evaluate how this market is likely to change over time.

For example, if 1000 patients *initiate treatment* each month, the numbers of patients on each treatment option produced by extended dynamic model including the lines of therapy progression is shown in Figure 10.

Figure 10 shows that while the number of patients on Treatment A is constant (1000 patients initiate treatment each month, but 1000 switch to Treatment C as

* In an extended model, patients might also start on Treatment A as they are reinitiating treatment. This dynamic is easy to incorporate, and does not change the fact that they are not reaching this treatment through a line of therapy progression.

per Table 5), the numbers of patients on other therapies is increasing rapidly. From a patient share standpoint, Treatment A is losing ground, even though it is a defined as an exclusive first-line therapy. In the battle to be the first entry point in a line of therapy progression, many pharmaceutical firms forget this one simple fact: *Patient shares and subsequent sales trajectories depend not only on how many patients are **acquired**, but also on how many patients are **retained***. Although positioning a product as first-line therapy can be important from a strategic standpoint, the ultimate performance of the drug will be the result of *all* the dynamics that affect the number of patients treated by that therapy. Acquiring patients early in their treatment progression is a good thing, all else being equal. But as this example shows, a "leaky bathtub" often results in competitors gaining from the patients captured and then lost, as quantified by a line of therapy analysis using data-driven patient dynamics.

Strategic Implications of Dynamic Modeling of Lines of Therapy

If the majority of patients progress through a defined progression of treatment options during the course of their therapy, the dynamics associated with treatment changes are best represented using an explicit lines of therapy approach. The resulting dynamic model simulation results can provide a Brand Planner with a much better picture of patient treatment dynamics and the strategic implications of various positions in a line of therapy progression. Some of our clients compete in markets where the first-line therapy is the only treatment that up to 50% of patients ever receive, making early entry into a line of therapy progression that is vital for new compound. In other indications, patients switch from early lines of therapy rather rapidly, making retention a much more important issue regardless of position in the treatment sequence. Assessing such scenarios through the use of Dynamic Modeling allows lines of therapy progressions to be rigorously analyzed, and an effective set of associated strategies to be tested.

We will examine the concept of using dynamic models for strategy testing and development in chapter 11, but even this simple extension of the Standard Template should give the reader some ideas about possible strategic intervention points. The visual nature of the stock/flow framework and the ability to integrate actual market-level data has led many of our clients to include lines of therapy structure in their simulation models of marketplace dynamics.

DISAGGREGATION

Dynamic Modeling software allows a great degree of flexibility in disaggregating stock/flow structure into more detailed components, as given in the example of disaggregation of patients according to treatment options in chapter 4. Disaggregation is a powerful way to "break out" a collected group of patients according to defined characteristics, and is often used to provide detailed results after an aggregate structure has been defined. In general, disaggregation involves first defining a

dimension—a broad class along which patients will be grouped. A list of common dimensions includes, but is not limited to, the following:

- age
- gender
- income level
- geography
- disease severity

Second, that dimension must be defined by **elements**—grouping choices that determine how the dimension will be disaggregated. Common elements for frequently used dimensions are as follows:

- young, middle-aged, and old
- male and female
- low, middle, and high income
- United States, Europe, and Asia
- mild, moderate, and severe

Dimensions have no conceptual upper limit regarding the number of elements to include, but practical constraints revolve around data availability and increasing model detail. For example, it is theoretically possible to build a dynamic model of the U.S. population consisting of roughly 300 million elements—one for each American citizen*. But such a model might be unwieldy and challenging to analyze. A more practical approach is to identify useful "groupings" and then expand the number of groups/elements to meet evolving strategic needs.

In our consulting practice, we have found it helpful to categorize potential disaggregation dimensions into two classes. **Arrays** are defined as disaggregations that can and often exhibit interelement movement. These dynamics are often important in modeling efforts to categorize patients moving or advancing through stages defined as elements of the array. For example, setting up a Disease Severity disaggregation as an Array would allow the dynamic model to track both the elements of severity (Mild, Moderate, and Severe) *and* the advancement of patients through those elements. ("How many Mild patients advance to the Moderate category each year?") **Segments** are defined as disaggregations where interelement movement is either theoretically impossible, possible but not easily quantifiable, or beyond the scope or intent of the modeling effort. Establishing an Income Level disaggregation as a Segment, for example, would allow the dynamic model to track people by a defined wealth metric, but *not* quantify or address interelement flows. ("How many low income citizens enter the middle class each year?")

Segments are generally easier to implement than arrays due to difficulty in obtaining good data around interelement dynamics. Examples of these two categories of disaggregation are shown in Table 6.

* This approach is more often utilized using what is called an agent- or individual-based approach (see chap. 9).

Table 6 Categories and Examples of Disaggregation Types

Arrays	Segments
Treatment Options	Race
Age	Religion
Disease Severity	Gender
Experience Level	Income
Seniority	Geography

Clearly, some of these disaggregation dimensions could be classed as either Arrays or Segments depending on the purposes of the model. Dynamic Modeling software technology allows this choice to be changed easily at any time in the modeling process.

When defining elements of a dimension disaggregation, the concept of a MECE set is worth revisiting. Introducing groupings or subcategories that "break down" an aggregate stock into component detail requires that those details can "sum up" to the aggregate number. An MECE set (see chap. 4) defines a disaggregation such that each patient is in one, and only one, element at a single time point. For example, a disaggregation along demographic lines with elements of U.S. citizens and naturalized citizens would allow overlap between the elements. A better designation would be native-born citizens and naturalized citizens, with the sum of the two accounting for all U.S. citizens.

Disaggregation introduces detail, but not necessarily dynamic complexity into the modeling process. In other words, disaggregating a model does not define a new model stock/flow structure, identify additional dynamic pathways for patient movement, or change the aggregate diagram as defined in the Standard Template or any extensions thereof. Instead, disaggregation simply takes broad patient categories (stocks) and associated dynamics (flows) and breaks them down into finer and more detailed subgroupings (some of which necessitate interelement movement). Disaggregation can often add value but with the following caveats:

- Disaggregation involves additional data cuts simply to populate initial values of stocks.
- Corresponding longitudinal analysis (see chap. 6) becomes increasingly difficult as sample sizes decrease due to increased disaggregation.
- A top-down approach—build the aggregate structure, disaggregate as data, and interest dictate—can be successful.
- A bottom-up approach—define all the dimensions/elements of interest, then try to build a model and gather data to support it—is a difficult path to success.
- Disaggregation detail is geometrically troublesome, particularly when dealing with interelement dynamics in an Array. Suppose a working team wants to set up six elements in an Array. The resulting interelement 6×6 matrix (36 elements) is more complicated than a 2×2 (4 elements) by a factor of 9.

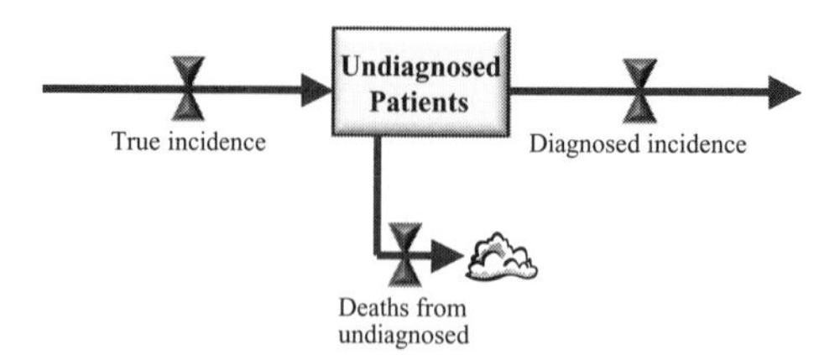

Figure 11 Aggregate stock and flows for undiagnosed patients.

- The greater the level of disaggregation, the greater the chance that the actual modeler will be the only person to use the model. The level of detail embodied in heavy disaggregated models is often overwhelming to potential end users.
- True model disaggregation should be made on the basis of following questions:
  - Are the parameters that govern behavior of the dimension/element significantly different from a strategic standpoint?
  - Can data be collected to support this hypothesis?
  - Will including this dimension/element make the model more usable or less usable?
  - Do model results based on the proposed disaggregation allow the team to make better strategy decisions?
  - Can the disaggregation be applied outside the model? In other words, can the aggregate stock/flow outputs be decomposed into constituent dimensions/elements after results are computed?

With these disclaimers in mind, note that disaggregation can be a powerful addition to any dynamic model, and is fairly easy to implement in currently available software packages. As an example, take the simple stock/flow piece from the Standard Template as shown in Figure 11.

Case Study: Disaggregating Stock/Flow Structure of Undiagnosed Patients

Suppose a Brand Planning team had information about the distribution of Undiagnosed Patients by gender and wanted to disaggregate the dynamic model across that dimension. Conceptually, this process would involve simply breaking down the stock as shown in Figure 12.

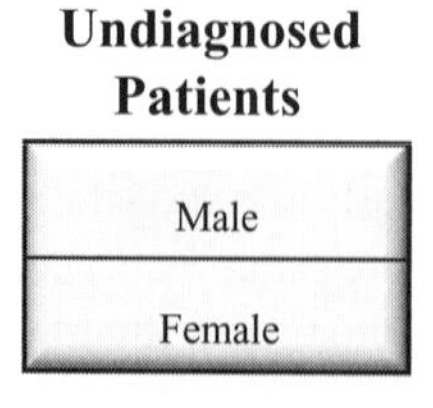

Figure 12 Undiagnosed patients stock disaggregated into gender dimension.

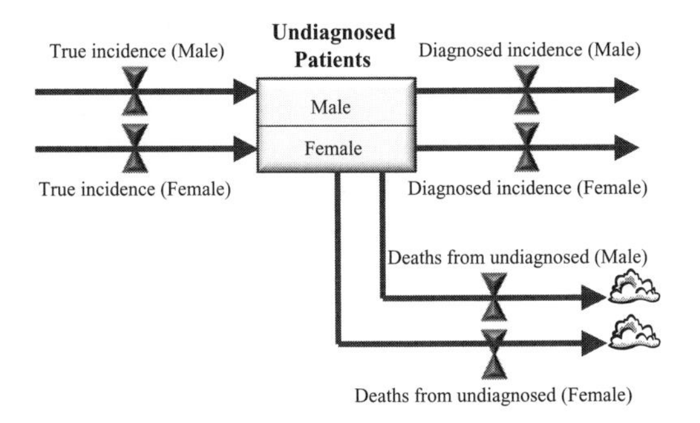

Figure 13 Undiagnosed patients and attached flows disaggregated into gender dimension.

As noted in chapter 4, the introduction of a dimension to stock requires that the associated flows are disaggregated in *exactly the same dimensions*. Conceptually, the diagram can be depicted as shown in Figure 13* .

In effect, the basic dynamic structure has been copied and then pasted to arrive at two identical stock/flow representations to track male and female patients separately. Gender-specific data on *true incidence, diagnosed incidence, and deaths from undiagnosed* could then be used to populate the above structure and produce outputs corresponding to each element of the dimension.

In this example, the gender disaggregation introduced a simple one-dimensional segment (gender) with two elements (male, female), but Dynamic Modeling software is very flexible in adding additional dimensions and/or elements of disaggregation, as shown in Figure 14.

Note the inclusion of Array dimensions (Age, Severity) necessitates interelement flows to be included. As a practical matter, these flows are shown as going only one way, although in theory severity flows could be bi-directional, such that Severe patients could return to being classed as Moderate. In addition, the Gender dimension has been identified as a Segment rather than an array, thus eliminating potential intergender flows with apologies to the plastic surgery community.

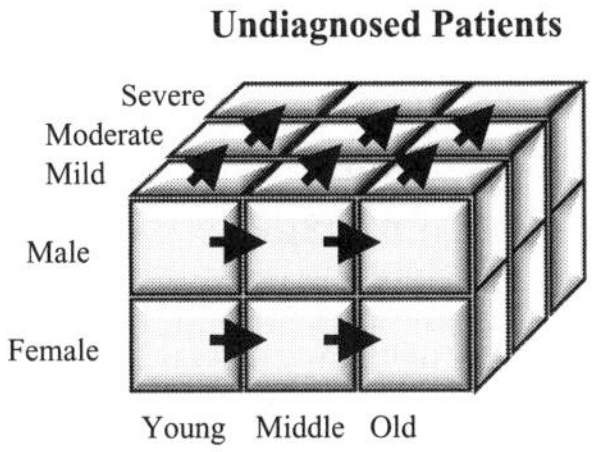

Figure 14 Three-dimensional disaggregation of undiagnosed patients stock by age, gender, and disease severity.

* Dynamic Modeling software packages generally do not show the precise detail of this expansion pictorially, and instead identify that a stock and its associated flows are disaggregated without explicitly showing the associated dimensions.

Disaggregation should be used sparingly and *only* after aggregate structure has been discussed, data sources identified, and initial aggregate simulation results have been interpreted. Highly disaggregated models have a tendency to be overwhelming in terms of data collection, ease of use, and transfer between team members. To paraphrase Albert Einstein, the best models are as simple as possible... but no simpler. Modeling teams should be aware of the cost/benefit tradeoffs before embarking on a detailed disaggregation exercise.

Strategic Implications of Dynamic Modeling Disaggregations

Increased availability of patient-level data has made disaggregating dynamic models a popular option for many of our clients, subject to the caveats listed in this chapter. Such extensions first identify and size potential market segments related to patients and physicians, which can often be useful in understanding the strategic importance of various groups. Dynamic Modeling can then capture and quantify the differential behavior of these constituent groups to discover how their activities vary. Identifying differential dynamics among key market segments can help unlock potential strategies for marketing leverage to inform the Brand Planning process.

Extensions of the Standard Template demonstrate the flexibility of the Dynamic Modeling approach, as they allow for the structure that addresses some of the complexities of individual disease marketplaces. The examples detailed in this chapter are by no means comprehensive, but should illustrate how this methodology can be adapted to meet specific situations that extend the Standard Template framework. Some of these extensions will be incorporated as the example dynamic model is expanded throughout the remainder of this work.

SUMMARY

- Standard Template is easily extendable to address more complex and detailed aspects seen in many pharmaceutical markets.
- Extensions do not change the overall approach or methodology of Dynamic Modeling, but simply enhance and augment the degree of analysis.
- Categories such as undiagnosed patients can be expanded to explore the subprocesses through which patients pass while in that stage.
- Nonchronic and/or episodic diseases can be analyzed using concepts of period versus lifetime prevalence, recover and relapse rates, and the dynamic epidemiology of periodic illnesses.
- Categories of common misdiagnosis can be included to detail potential pathways to correct diagnosis.
- Lines of therapy progression can be incorporated to visually depict and explicitly quantify the sequential treatment patterns of patients.

- Disaggregations can be useful in breaking aggregate stock/flow structure into more detailed subcategories, but can exponentially increase the complexity of the Dynamic Modeling framework.
- Potential model extensions should be carefully evaluated to ensure that the resulting model is still usable and not overwhelmingly detailed.

6

Data for the Standard Template Dynamic Model

INTRODUCTION

This chapter will cover the following topics:

- Required patient stock data
- Sources of stocks data in the Standard Template
- Required Patient Flow data
- Longitudinal integrated claims databases
- Analysis of Treatment Change Matrices

One of our clients describes dynamic models as being like chocolate chip cookies—strange as that may sound. Both are visually appealing, but the best results come when quality ingredients are used and care is taken to measure and add those ingredients appropriately. Good data are the flour, sugar, vanilla, and chocolate for a dynamic model, and this chapter will show how key data ingredients are often found, collected, measured, and incorporated. And like a great batch of warm cookies, the results of applying a solid data foundation to the Standard Template are often more than the sum of the parts. (Please excuse us if you feel this analogy is "half baked.")

The old saying of "garbage in, garbage out" certainly applies to all types of modeling methodologies, but dynamic models do have one advantage over certain other approaches when it comes to data. Because dynamic models explicitly account for the relationships between accumulations (stocks) and movements (flows), they have a built-in "sanity check" for model results. As we discussed in chapter 3, the operational nature of dynamic models helps support internal data consistency. In addition, chapter 13 will focus on how the stock/flow structure of dynamic models places bounds on the possible results of a modeling effort. The central tenet of the Dynamic Modeling methodology is that the stocks can only change as a result of their associated flows, so integrating data from each of these fronts provides a fundamental cross-check to prevent data incompatibility. While dynamic models are not immune to the problems plaguing all types of projection methodologies, in our experience they provide a smaller margin of error due to their explicit operational nature.

The good news for any potential Dynamic Modeling effort is the wide availability of solid data, particularly on the patient level, that has been developed over the last decade. The advent of large, representative patient databases has made the data collection process more robust, more applicable to specific disease indications, more accepted by potential decision makers, and more insightful when applied to specific business questions through the framework of an integrated dynamic model. Patient-level data for Europe still trails that of the United States both in depth and breadth, although the establishment and acceptance of the European Union is slowly changing this situation. Sound patient-level data and a solid methodology for collecting and analyzing it can impact strategic decisions for a compound, given the importance of patient behavior in determining overall marketplace dynamics. A common misconception is that the advent and wide availability of patient-level data means that the pharmaceutical industry is in the Information Age. Because information on patient behavior is readily available and relatively inexpensive to collect, store, and analyze for all competitors, the period of information for the sake of information has long since passed. Instead, pharmaceutical firms now find themselves in the intelligence age—a stage in which the ability to use data more effectively than the competition is of vital strategic importance. And it has been explained throughout the course of this work that the Dynamic Modeling methodology provides an explicit way to integrate widely available data into a consistent and operational framework.

PATIENT STOCK DATA

The types of data needed to populate dynamic models are as varied as the varieties of the models themselves, but some examples of data collection and sources from the Standard Template should help illustrate the data-gathering process. The Standard Template establishes some precise definitions related to its associated stock/flow structure. Populating this structure with valid data involves corresponding rigor. Initial estimates of model parameters are often revisited and revised over the course of a modeling project, and are subject to sensitivity analysis in the form of Monte Carlo or similar multivariate simulations. However, the Dynamic Modeling process requires some initial patient data estimates as a starting point.

As we described in chapter 3, dynamic models must begin with initial values of each stock in the model structure.* To do so, an effective starting time for the model must be determined and then relevant data pertaining to that time point must be gathered. The starting point for the Standard Template is usually at the

* Initial values of stocks can be estimated if reliable data exists regarding the inflow pattern and the average length of time items spend in the stock. (see chap. 7).

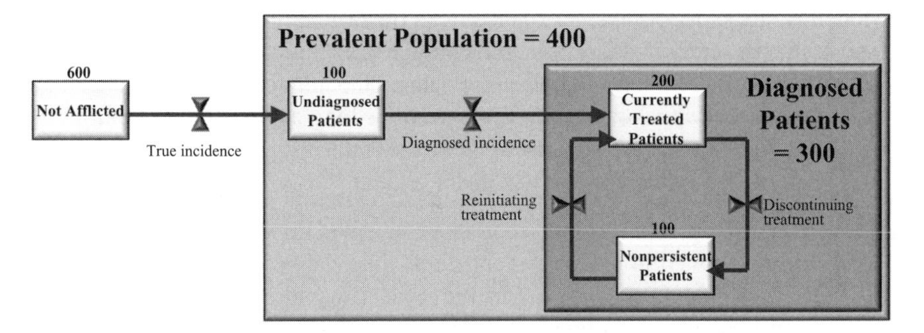

Figure 1 Example Standard Template populated with initial stock values (January 2009).

beginning of a particular calendar year, and data regarding the number of patients in each of the model stocks can be incorporated into the model diagram as shown in Figure 1.

Recall from chapter 4 that the Currently Treated Patients stock is often disaggregated to include treatment options in an MECE set. This set is designed by a working team to capture currently available products and/or regimens and those entering the market within the time horizon of the model. Initial numbers of patients on each of these treatment options need to be collected, and can be depicted as shown in Figure 2.

Recall that the values in Figures 1 and 2 represent only *initial* numbers of patients in the Standard Template. Subsequent stock values will be the result of the dynamics of the system, and the magnitude of inflows and outflows will determine how the numbers of patients in each stock will change over time. The dynamic model will make those calculations automatically through the course of a simulation, but establishing a solid starting point is an important step in the overall modeling process. Our clients find that the process of determining appropriate estimates for the initial numbers of patients in the various Dynamic

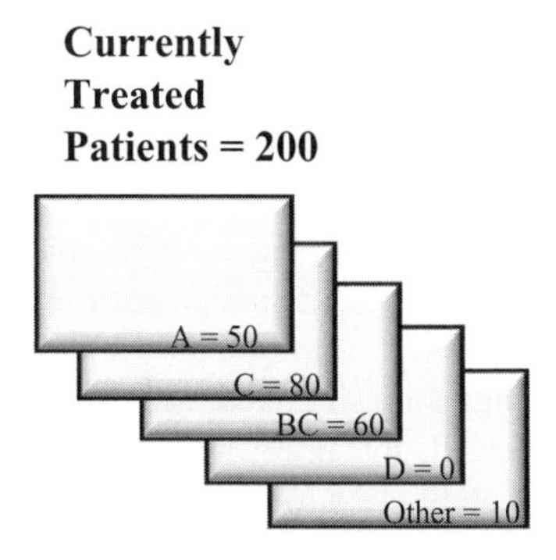

Figure 2 Example initial MECE set stock values (January 2009).

Modeling stocks is a valuable exercise in its own right, often providing strategic insight into the current status of an indication marketplace.

In general, defining and collecting data on the initial numbers of patients in each model stock is relatively straightforward as long as the strict definitions of the Standard Template are adhered to. Note the following:

- In total, the stocks in the Standard Template represent the entire patient population in a geographic region. The same concept of an MECE set (see chap. 4) should be applied to all the patient stocks in a defined dynamic model. The MECE designation allows the dynamic model to capture each patient once, and only once, at any given time point. In many indications, the pool of potential patients is age specific, so the universe of "population" is often restricted to a particular age demographic. Within that age category, however, the principle of complete patient inclusion still applies.
- The MECE treatment option set in the Currently Treated Patients stock provides a similar grouping algorithm to ensure that all treated patients are represented but in a manner that prevents double counting.
- The Non-Persistent stock represents patients who are not on *any* treatment at a particular time point. The criterion for this designation often varies by indication. For example, in some markets such as rheumatoid arthritis, where monthly injections are common therapy, a patient might be labeled Non-Persistent by missing only one month of treatment. In other asymptomatic and/or chronic indications such as hypertension or dislipidemia, the Non-Persistent designation might occur only if a patient has not been on any form of medication for three to four months. Regardless of the persistency threshold, the definition of Non-Persistent is vital in correctly capturing key patient dynamics of therapy discontinuation and reinitiation in the Standard Template.

As was discussed in chapter 5, introducing disaggregations into the Standard Template requires additional detail in establishing initial numbers of patients in each stock. For example, disaggregating the model to account for gender would require initial patient data on that dimension, as shown in Table 1.

Disaggregating the Standard Template on dimensions such as age, disease severity, and/or comorbidities is also possible, but the success of these model expansions depends on data availability and quality. Dynamic Modeling software imposes no upper limit to the dimensions of disaggregation, but practical limitations are soon met with increasing detail complexity. **A good rule of thumb**: *If initial stock values are difficult to collect due to excessive disaggregation, the corresponding data related to the associated flows across those same dimensions will be impossible to find.*

The Standard Template often varies by the indication it is addressing, the detail complexity of the dynamic model underlying it, the time frame it runs through, or even the specific strategic questions it is attempting to answer. But for all such variations, one basic condition holds true: The Standard Template dynamic model must begin with initial numbers of patients for every stock at a particular

Table 1 Example Initial Stock Values by Gender (January 2009)

Variable	Total	Male	Female
Population	1000	500	500
Not afflicted	600	300	300
Prevalence	400	150	250
Undiagnosed	100	40	60
Diagnosed	300	160	140
Nonpersistent	100	50	50
Currently treated	200	110	90
A	50	30	20
C	80	35	45
D	0	0	0
BC	60	40	20
Other	10	5	5

starting point. Without these initial conditions firmly established and backed by quality data, subsequent analysis of market dynamics will lack a sufficiently firm footing.

SOURCES OF PATIENT STOCK DATA

The data needed to establish the initial patient stocks can come from a variety of sources, many of which are readily available to most pharmaceutical firms. For many of our clients, these data sources have already been utilized to develop static or spreadsheet models for forecasting purposes, and these efforts are often a useful starting point. It is important to note, however, that the Dynamic Modeling methodology requires a level of rigor that may not have driven existing data collection efforts. Care must be taken before simply "plugging in" patient data from other approaches, as the outputs of the resulting Standard Template model rely on precise definitions and quality data inputs.

Many of our pharmaceutical clients employ in-house epidemiologists to help with quantifying the required patient numbers for the Dynamic Modeling approach. Epidemiologists often help inform the working team of data availability, possible segmentations, and evolving disease metrics that might impact an indication marketplace. If such internal resources are not available, we have found the following to be good sources of patient-level data to populate the stocks of the Standard Template:

- Population—The U.S. Census Bureau Web site (www. census.gov) posts both historical and projected demographic data cut in a variety of different dimensions. These projections are also available through the Center for Disease Control (www. cdc.gov) and are updated periodically.

- Prevalence and diagnosis—A variety of published literature examines these disease-specific data, drawing on peer-reviewed research studies and often forecast over some time interval. Advocacy groups or foundations focusing on specific indications usually provide information on disease epidemiology, usually through their respective Web sites. The CDC is again a good starting point for aggregate epidemiology estimates, as the site has a searchable database of published literature regarding certain disease areas. The National Center for Health Statistics (NCHS) is a CDC division that collects and publishes cross-sectional survey data in the form of the National Health and Nutrition Examination Survey (NHANES) report. Various versions of NHANES data often provide a starting point for epidemiology projections, but our clients increasingly question such estimates due to delay between data collection, analysis, and ultimately publication. More timely epidemiology is available through companies such as Episource, Decision Resources, and Data Monitor that sell consolidated reports on key epidemiology metrics in selected countries, usually the United States, Europe, and Japan. Estimates on epidemiology often vary widely, and an important part of the Dynamic Modeling process is to agree on both the definitions and the associated parameters to be used in the project. Some of our clients choose to enlist the help of independent professional epidemiologists to add a degree of rigor to these analyses or to provide patient projections on dimensions not included in syndicated epidemiology reports.
- Currently treated/nonpersistent—Syndicated reports often estimate the treatment rate for diagnosed patients, although the definitions of this metric must be carefully analyzed to ensure that the data fits into the Standard Template framework. Online resources such as PubMed provide a searchable repository of published research in various disease indications that may provide insight into current treatment rates. Pharmacy claims databases often provide the best sources for determining patient distribution across various treatment options, including patients designated as Non-Persistent. Longitudinal forms of these sources allow for initial estimates to be confirmed based on dynamic model results. Other sources for patient distribution by treatment include either primary or secondary market research on patient records, physician charts, and medical records databases. In our experience, however, pharmacy claims provide the best representation of what *actually* happens in the marketplace, as they capture actual patient behavior—as opposed to physician intentions, for example.
- Other stocks/extensions of Standard Template—Modifying and/or extending the Standard Template requires additional data regarding initial numbers of patients in each of the introduced stocks. Data sources to populate this additional structure are evaluated on a case-by-case basis, and care must be taken to ensure that a data provider has sufficient sample size and the ability to provide data in the appropriate level of patient detail to satisfy the needs of the Brand Planning team. We have often worked with teams wanting a very complex

model structure for which data is not readily available. A good question to ask in such a situation is "If this extended stock/flow structure is so important, why is data associated with it nonexistent?" Such introspection can often help guide the Dynamic Modeling process and prevent the project from becoming too detailed without associated data to support it.

Strategic Implications of Obtaining Valid Patient Stock Data

If you do not know where you are, it is hard to figure out where you can get to, so assessing the current state of the system correctly is a vital part of any market evaluation process. In many cases, the Dynamic Modeling methodology takes probabilities of patient dynamics (developing a disease, progressing through various treatment options, discontinuing treatment, etc.) and applies them to aggregate epidemiology numbers to arrive at calculated flows of patients in a marketplace. As such, the initial conditions of the disease marketplace are crucial to understanding the dynamics from which effective marketing strategies can be developed. In addition, simple sizing of patient groupings or detailed segmentation analysis may provide strategic insight into stocks of patients representing potential marketplace opportunities. Initial stock values by themselves are not dynamic, but do provide the basis from which the Dynamic Modeling approach can be applied.

Establishing good data to populate the initial values of the Standard Template stocks may seem like a trivial task, but the process often involves estimation of parameters and triangulation among various sources. This exercise is both necessary for the completion of the dynamic model and valuable in its own right, however, as understanding the relative strengths and weaknesses of various data can translate into a clearer picture of a disease marketplace and its strategic implications.

PATIENT FLOW DATA

Operationally, it is the flows of patients, rather than the stocks to which they are attached, that are a primary driving factor in how pharmaceutical markets change over time. From an epidemiology standpoint, the movements of patients into a disease category through incidence or out of it through deaths or recovery dictate how the prevalence in an indication is likely to evolve. Chapters 4 and 5 addressed how these flows of patients could be derived from a combination of epidemiology forecasts, mortality information, and recovery/relapse rates, and need not be revisited here.

But from a product standpoint, the flows associated with patients and their pharmaceutical treatment are what drive prescriptions and ultimately revenues in an indication marketplace. These key treatment dynamics are the driving forces behind changing market share and patient share over time, and are captured very operationally in the Standard Template dynamic model. Chapter 4 depicted how these key treatment dynamics can be categorized and expressed as flows of patients, which are as follows:

- initiating treatment (diagnosed incidence)
- changing treatment
- discontinuing treatment
- reinitiating treatment

Recall that these patient flows are often disaggregated to allow for further behavioral details regarding key patient segments. For example, a disaggregation across an age dimension would allow for varying patient flow treatment dynamics according to patient age. Corresponding data sources can be queried along the same dimensions of disaggregation to provide segment-specific data.

Fifteen years ago, quantifying such treatment change dynamics was extremely difficult and often relied on surveys of physician intentions or complex forecast algorithms, which could not be supported by patient-level data. The advent and availability of patient databases has proven to be a gold mine for the Dynamic Modeling process, and has transformed the data collection and validation process into a rigorous and enlightening exercise.

Without question, the best sources of data to populate the patient flow treatment dynamics in the Standard Template are **longitudinal integrated claims (LIC)** databases. These databases are available from a variety of third party providers, and many of our clients subscribe to their services for the purposes of collecting patient-level data to populate dynamic models. LICs have the unique advantage of matching a pharmacy claim with a disease diagnosis at the individual patient level, ensuring that the resulting analysis is indication specific. LIC databases establish individual patient records* that have a number of useful characteristics when it comes to capturing patient flow treatment dynamics:

- *Longitudinal*—Patients are tracked through time, so their behavior over a period of months or even years may be collected and analyzed.
- *Integrated*—Patients are diagnosed and coded according to the International Classification of Diseases (ICD) standards.
- *Claims*—Pharmacy claims are recorded, allowing patients' treatment regimen patterns to be collected and analyzed.

For indications in which treatments are not specific to a particular disease, integrating a diagnosis with a prescription is vital. Antipsychotics are indicated for a variety of mental illnesses including schizophrenia, bipolar disorder, and even depression and so simply examining prescription data does not in itself reveal the disease for which any particular antipsychotic was prescribed. But by matching a diagnosis with a pharmacy claim, LIC databases can ensure that *only* schizophrenia patients are being tracked, for example, not simply all patients

* Although records are created for individual patients, the data from LICs is aggregated according to defined characteristics and patient confidentiality is never compromised.

receiving an antipsychotic. Such detailed information, coupled with the ability to track groups of patients over time, make LIC databases the data source of choice for populating patient flow data in Dynamic Modeling efforts.

A number of vendors such as MediPlus® from IMS Health, MarketScan® from MedStat, Caremark Rx, PharMetrics, Verispan from Scott-Levin, Surveillance Data, Inc. (SDI), Synovate, and GE HealthCare specialize in these types of LIC databases, and their relative merits for a particular indication should be evaluated by team members to ensure large, representative, and nonskewed samples from which to draw. Many pharmaceutical companies already purchase data from these sources or have in-house resources to evaluate the strengths and weaknesses of each.

Such LIC databases provide a wealth of information regarding patient flow treatment dynamics. Their longitudinal nature allows the dynamics of treatment changes to be calculated, and the fact that their pharmacy claims can be tied to specific diagnosis codes ensures that the analysis is being applied to groups of patients in a particular indication. These characteristics are crucial in developing a good dataset to populate the Standard Template dynamic model.

In our years of consulting in the pharmaceutical world, we have often seen key patient treatment dynamics expressed longitudinally as a persistency curve. An example of this type of analysis is shown in Figure 3.

As was discussed in chapter 4, the typical persistency curve as seen in Figure 3 captures drug-specific persistence, but does not designate or capture the true operational dynamics of patients switching treatment or discontinuing all type of prescription medicine for their condition. While illustrative, typical persistency curves often mask some key patient dynamics that can be of strategic importance.

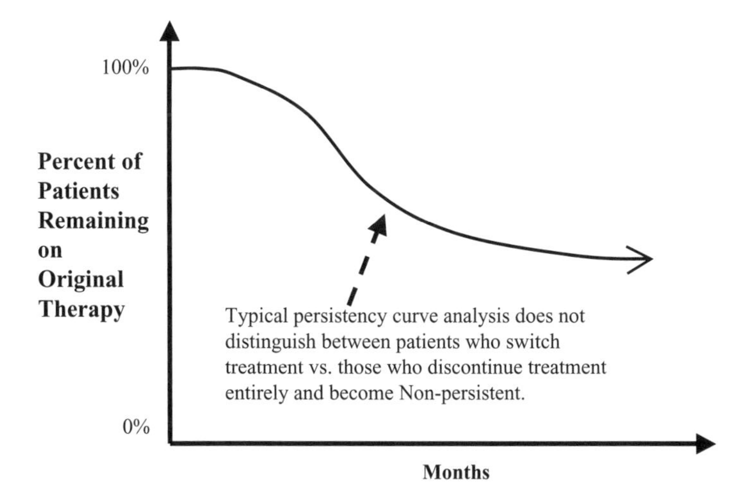

Figure 3 Typical analysis of persistency in pharmaceutical industry.

Table 2 Example Single Patient Treatment Dynamics (January–December 2009)

Treatment	Jan	Feb	Mar	Apr	May	June	July	Aug	Sept	Oct	Nov	Dec
A	•	•										
C			•							•		•
BC				•	•	•						
Other											•	
No Rx							•	•	•			

In contrast, LIC databases provide the means to quantify a variety of types of treatment dynamics in an indication. Imagine a database query that indexes a patient at a particular time—January of 2009, for example. The query can establish parameters for inclusion—a particular indication, for example, or only patients older than the age of 20. In addition, the MECE treatment option set in our continuing example can be integrated into the analysis to ensure that data results map directly to the defined Standard Template structure. The LIC in this example query provides the means to track this same patient over the course of next 12 months to see what happens to the patient's treatment regimens, as illustrated in Table 2.

The results of the LIC database query shown in Table 2 detail this patient's treatment regimen coverage during each month of 2009.* The longitudinal data for this example patient shows a progression through Treatments A, C, and BC before discontinuing therapy altogether and being categorized as having No Rx. After three months of no prescriptions for this particular indication, the patient reinitiates treatment on a regimen in the Treatment C category, jumps down to Treatment Other for one month, then switches back to Treatment C.

Clearly, the patient in this example shows a high number of therapy changes in the form of switches, discontinuations, and reinitiations over the course of the year. It is difficult to intuit overall market behavior from a single sample such as this, as an individual patient's propensity to change treatments may or may not be representative of the entire patient population.† However, individual data can be aggregated to achieve sufficient sample sizes of cohort groups of patients, which in turn can be analyzed to provide data to populate the Standard Template dynamic model.

Such an LIC analysis can be used to determine the likelihood or probability of patients moving between model stocks in each period of time. For treatment change dynamics, this translates to changing treatment options or moving into or

* Treatment changes occurring within a given month are generally not captured explicitly in an LIC, but can be estimated through analysis of other treatment change metrics and the degree of combination usage in a given indication.

† Truly new patients, for example, are often more prone to treatment changes than established patients. The designation of new versus continuing patients is often included in the database query to get data results specific to each group.

Table 3 Example Treatment Change Probability Matrix Including Nonpersistent (2009)

From/To	A	C	BC	Other	No Rx	Total
A	N/A	1%	3%	3%	4%	11%
C	4%	N/A	4%	4%	4%	16%
BC	0%	5%	N/A	10%	15%	30%
Other	2%	5%	2%	N/A	5%	14%
No Rx	10%	0%	10%	0%	N/A	20%

out of the Non-Persistent category. The Standard Template model can then apply those derived dynamic probabilities to the overall number of patients in each stock to determine the magnitude of the flows associated with it.

For example, assume patients from an LIC database were analyzed and aggregated to determine their likelihood of moving between treatment options in a particular marketplace. This analysis can be shown as a 2-dimensional treatment change matrix (see chap. 4) consisting of switching FROM a particular therapy option TO any other therapy option in an average month. In this case, the No Rx category can be thought of as an option to not be treated, representing the Non-Persistent stock of patients. The results of this analysis are shown in Table 3.

First, note the column titled TOTAL in Table 3. This metric represents, in aggregate, the probability that a patient who begins a given month on given treatment option will be on another Treatment choice (including No Rx) by the next month. For example, 11% of Treatment A patients will switch to another drug or drop off treatment entirely in an average month. Of those 11% changing treatment, 1% will go to Treatment C, 3% will go to Treatment BC, 3% will go to Treatment Other, and 4% will completely drop off treatment and be classified as having No Rx (representing Non-Persistent.)

Second, note how the aggregate switching probabilities in Table 3 begin to give insight into market dynamics. Treatment BC shows very high treatment change rates (30% per month) indicating market dissatisfaction with something about the regimen's efficacy, safety, side effect, or even cost.

Finally, note these treatment change probabilities in Table 3 can give indications regarding lines of therapy, as was shown in chapter 5. For example, note that 0% of patients switch from Treatment BC to Treatment A. Similarly, all patients reinitiating treatment (coming from the No Rx category) go on either Treatment A or Treatment BC, but never reinitiate to Treatment C or treatment other. Such analysis of the Treatment Change Matrix begins to show common pathways for patients as they progress through various types of treatment.

Dynamic models take this type of probability data and incorporate a time element to make monthly calculations regarding flows of patients. For example, given initial conditions regarding the number of patients on each treatment option, the Standard Template model can produce the actual movements of patients between treatment options over the course of a given month, as shown in Table 4.

Table 4 Example Patient Flow Treatment Change Matrix (2009)

Initial no. of patients	From/To	A	C	BC	Other	No Rx	Total
50	A	N/A	0.5	1.5	1.5	2.0	5.5
80	C	3.2	N/A	3.2	3.2	3.2	12.8
60	BC	0.0	3.0	N/A	5.0	9.0	18.0
10	Other	0.2	0.5	0.2	N/A	0.5	1.4
100	No Rx	10.0	0.0	10.0	0.0	N/A	20.0
	Total	13.4	4.0	14.9	10.7	14.7	

Note that the TOTAL column in Table 4 shows the entire number of patients moving FROM a given treatment option—5.5 treatment A patients move on to another treatment option in January, for example. The TOTAL row shows the entire number of patients moving TO a given treatment option—13.4 patients go to Treatment A over the course of the month. The difference between these two outputs (13.4–5.5 = 7.9) represents the net gain/loss of patients associated with that particular therapy option in the indicated month. The Standard Template dynamic model integrates these concepts of patient gain versus patient loss on a monthly basis, so that at the beginning of February 2009, Treatment A will have 57.9 patients—initial value of 50 plus the net gain of 7.9 patients in January. In February, 11% of the 57.9 Treatment A patients will have a treatment change, and the corresponding probabilities will be applied to the updated numbers of patients on each treatment option. Integrating the dynamics in this way produces the changing numbers of patients on any of the defined treatment options, including Non-Persistent.

The Advantages of Longitudinal, Patient-Level Integrated Data

Steven Bloom
Vice President, Sales
PharMetrics, LLC

Integrated medical and pharmaceutical claims provide pharmaceutical executives with a picture of patient treatment patterns across a wide range of health care issues, and the PharMetrics database containing de-identified pharmaceutical and medical records is the largest and most complete of its kind. This health insurance claims database is contributed by more than 70 geographically diverse payor organizations and captures the entire continuum of care for more than 57 million covered lives. The database is representative of the commercially insured population in the United States and contains information on all types of interactions a patient has with the healthcare system, including office visits, diagnoses, hospital stays, prescription treatments, and diagnostic procedures and tests.

Integrated claims databases, especially of the longitudinal variety, can be broadly leveraged by pharmaceutical organizations to support analyses across therapeutic areas and functions. For example, PharMetrics' patient-centric data can be used to support the efforts of Brand Planning, market research, pharmacoeconomics, safety, epidemiology, forecasting, and lifecycle management. This type of data can be used in a variety of analyses:

- Developing and refining market segmentations
- Informing brand positioning
- Illustrating patterns and pathways of treatment
- Identifying and valuing market opportunities
- Assessing the cost impact of treatment interventions
- Analyzing performance vis-à-vis the competition
- Informing forecasting models and primary research

Many of our clients apply the information gleaned from a customized examination of the PharMetrics database to inform and populate their Dynamic Modeling efforts. The large sample sizes available from this comprehensive data source provide statistical rigor to the analysis of patient behavior over time, giving a more accurate assessment of what truly happens in disease marketplaces. These treatment dynamics can then be applied to overall epidemiology numbers to arrive at a full understanding of the magnitude of flows of patients in a given therapeutic area. By leveraging the Dynamic Modeling approach with PharMetrics' wealth of indication-specific longitudinal data, teams are able to have a much clearer and more robust understanding of the actual patient and treatment dynamics driving the behavior of a particular disease area. From a Brand Planning perspective, such market insight is vital in evaluating the commercial potential of, and developing effective strategies for, compounds entering and competing in pharmaceutical marketplaces.

As an industry leader in providing patient-centric market intelligence solutions, PharMetrics as a long history of helping pharmaceutical, biotechnology, and medical device companies make better business decisions and improve brand performance by analyzing the information contained in our comprehensive database. Utilizing this information within the Dynamic Modeling framework represents the state-of-the-art integration between representative patient-level data and powerful new simulation tools. The results of this union continue to provide unique strategic insights for our clients across a wide range of therapeutic areas and business issues.

Treatment Change Matrix Dynamics

For simplicity, we have shown the Treatment Change Matrices as set constants that do not change over time. Increasingly, we advocate that our clients obtain at least three different "cuts" of historical data to determine if these patient movement probabilities are indeed time invariant. This more thorough look backward can

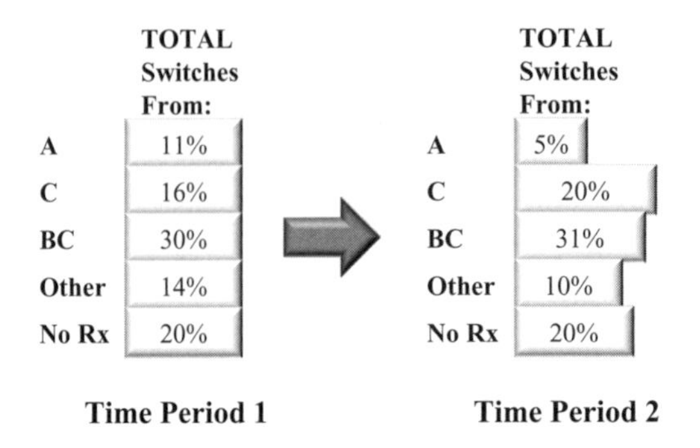

Figure 4 Graphical comparison of changing monthly treatment change probabilities.

give insight into how these dynamics might change going forward. (In a fully expanded Standard Template model use for projection purposes, these treatment change probabilities can be functions of other variables in the dynamic model, and thus change in response to changing conditions in the marketplace, new market entrants, or as the results of marketing strategies on the part of pharmaceutical firms. Chapter 10 will demonstrate the process of integrating various other causal factors into the patient flow Dynamic Modeling framework and its associated input parameters.)

For example, suppose that an investigation into an appropriate LIC database divided the historical patient behavioral data into two (for simplicity) relevant time periods. The identical methodology to determine each time period's Treatment Change Matrix can be applied, and the differences in results (if any) are likely to give an indication as to the magnitude and direction of key marketplace dynamics. Figure 4 shows graphically how one column* of results from a time-comparative analysis of varying Treatment Change Matrix results might be displayed.

Analyzing various "cuts" of data in this manner can give some very interesting insights into evolving patient behavior over time. Note that Figure 4 shows a clear increase in patient retention for Treatment A over time. In fact, these example data indicate that by Time Period 2, the rate at which Treatment A patients are exiting that therapy option has decreased roughly by half. Patients on Treatment C are experiencing the reverse dynamic, i.e., their exit rates are increasing from 16% in Time Period 1 to 20% in Time Period 2. Figure 4 also shows that the rate at which Non-Persistent Patients are reinitiating treatment is not changing over time; in each relevant time period, only 20% of the No Rx patients "switch" to become treated patients in any given month. In this manner, even before integrating this type of data into a fully functional simulation model, comparison of patient movements in separate historical time periods can begin to give marketing teams

* Typically, the results of the *entire* Treatment Change Matrix are compared in this fashion. Differences in results can be analyzed with statistical packages such as SAS or SPSS to determine the statistical significance (if any) between varying patient movement probabilities over time.

some very valuable insight into the dynamics, which are shaping the evolving marketplace.

Increasing the Detail of Treatment Change Parameters

For the sake of readability and continuity, the data parameters in our extended example have been set and unchanging; the probability of a patient moving from Treatment A to Treatment C in a given month, for example, was X%, or the average time that a patient would remain on a therapy before dropping off to become Non-Persistent was X months. Initial data estimates for such behavioral parameters often take this form that mathematicians would call deterministic (meaning that since the inputs to the model are always the same, the outputs will be the same as well.) However, the increased depth and breadth of patient level databases, as well as the growing sophistication of analysis and simulation tools to leverage it, allows for more detailed estimates of patient behavior to be including in a Dynamic Modeling context.

Patient Segmentation and Associated Data Collection

One way to accomplish this more detailed assessment of patient behavior is to disaggregate a stock/flow model into relevant Segments or Arrays (see chap. 5.) In this manner, the data analysis and associated model could look at and ultimately track "subgroups" of treated patients and assign group-specific behavioral parameters*. For example, instead of clumping all Treatment A patients into one aggregate category for analysis, the database could be queried to look at Treatment A patients meeting certain criteria—gender, age, ethnicity, geographic region, etc. Then the relevant Treatment Change Matrix would be derived for, and ultimately applied to, these specific groups in a disaggregated dynamic model. This sort of disaggregation has a number of advantages, such as

- subgroups can be defined based on patient characteristics in the associated patient database;
- groupings are often informed by marketing team decisions regarding important patient segments, both now and in the future;
- data results are easy to analyze in meaningful ways;
- categories can be "rolled up" to show aggregate results or "broken down" into component parts; and
- associated simulation results allow for "what if" analysis on specific patient disaggregation categories.

The caveat to significant model disaggregation involves both the data-collection task as well as the complexity of the resulting data analysis and dynamic model outputs. Recall from chapter 5 that model disaggregation results in a geometric increase in the number of parameters that must be determined from a

* See chapter 9 for yet another approach to disaggregation called "agent-based modeling."

detailed data analysis. For example, suppose a marketing team wanted to look at groups of patients along lines of gender (two elements), age (three elements), and disease severity (three elements), couple those groupings with the existing five treatment options in our continuing example (four existing therapy regimen as well as the No Rx category) and the number of patient segments for which parameter estimates must be made climbs rapidly. In fact, the expansion is multiplicative—5 (treatments) × 2 (genders) × 3 (ages) × 3 (disease severities) = 90. Collecting data on that many patient groupings is indeed possible, particularly for single point-in-time "slices," but incorporating longitudinal analysis for 90 (in this case) groups of patients often results in insufficient sample sizes. Our advise to clients is to always "do the math" on any proposed disaggregation protocol vis-à-vis the available patient sample from an appropriate database. Doing so early in the process is a much more efficient way to arrive at an effective dynamic model.

Probabilistic Treatment Change Parameters

A second method for departing from aggregate, unchanging parameters related to patient behavior is to enter the world of probabilistic or stochastic input variables. In a probabilistic model, at least one (and perhaps more) of the input assumptions is pulled from some sort of random distribution, such that the calculations within the simulation will be different each time the model is run because of changing input values. Therefore, the model does not have a single set of results, rather a distribution of potential outcomes given variation in individual input parameters*.

Introducing randomness into a dynamic model is a relatively easy process, but one that should be done with caution. Best practice is to build, test, revise, and finalize a model with deterministic inputs before embedding any stochastic parameters within it. In our many years of consulting, we have seen too many probabilistic models whose "interesting results" were simply due to formulation errors within the simulation environment. However, business people in the pharmaceutical world and beyond are increasingly comfortable with techniques such as Monte Carlo analysis and sensitivity results and so incorporating these concepts within a Dynamic Modeling context can be a valuable addition.

For example, assume for a moment that a marketing team is interested in a detailed look at treatment-specific persistency. In other words, how long do first-time patients stay on a particular therapy before switching to another drug? An analysis of an appropriately representative LIC database could identify a cohort of patients beginning therapy on a particularly treatment (Treatment X) and then track their subsequent prescription behavior through time to determine their treatment persistency. (This analysis would mirror the evaluation done on an individual patient as show in Table 2, but the results would be aggregated to the group level.) These patients could be categorized into different groupings based on their length of therapy, as shown in Figure 5.

* For more on probabilistic models in the context of modeling R&D pipelines, (see chap. 16).

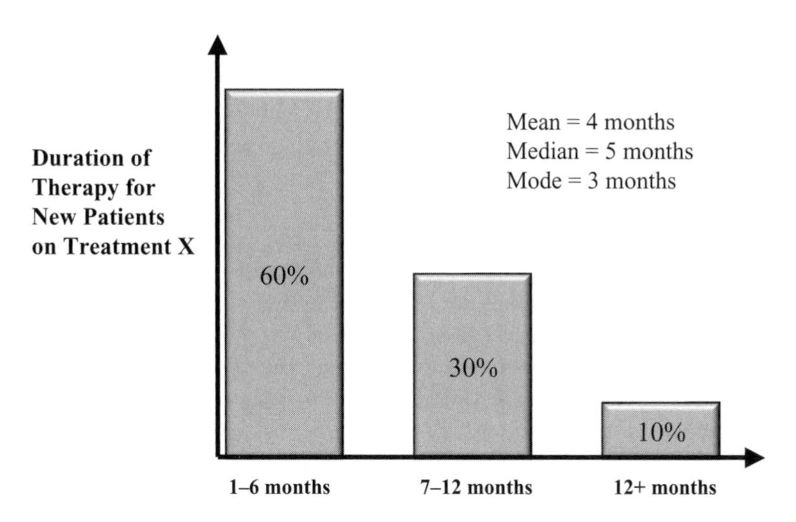

Figure 5 Example duration of therapy distribution for new patients on treatment X.

From a Dynamic Modeling standpoint, a very simple option would be to simply take the mean duration of therapy (four months) and use that as the input to a deterministic model. Another approach would be to create segments (see chap. 5 and section "Patient Segmentation and Associated Data Collection" in chap. 6) of patients based on the therapy duration groupings shown in Figure 5 and have the behavior of each individual patient segment be dictated by the mean therapy duration for that grouping. A third approach is to allow the duration of therapy to take on a range of values, and to use randomness to determine a range of possible model outcomes given uncertainty regarding this key input assumption*.

Figure 5 implies a distribution of Treatment X therapy durations for patients, and this variation may result from a number of factors related to the use of Treatment X, which are side effects, efficacy, safety, financial considerations, etc. In addition, this distribution implies something very important from a modeling perspective—that the marketing team cannot say exactly how long patients will remain on Treatment X before discontinuing use of the drug. While frustrating from the standpoint of wanting a definitive market projection, this uncertainty can be embraced by including variability in the possible therapy durations for patients on Treatment X. Incorporating these varying probabilities into a dynamic model will allow the team to determine the range of possible outcomes (such as the number of patients on Treatment X) over time.

First, a very simple dynamic model can be established to look at the stock of patients on treatment X and its associated flows (*starting Treatment X and leaving Treatment X*), as shown in Figure 6. Next, randomness must be introduced to

* A fourth technique, in which individual patients are given specific therapy durations based on parameter draws from the distribution shown in Figure 5 of this chapter, employs an approach called agent-based modeling and is detailed in chapter 9.

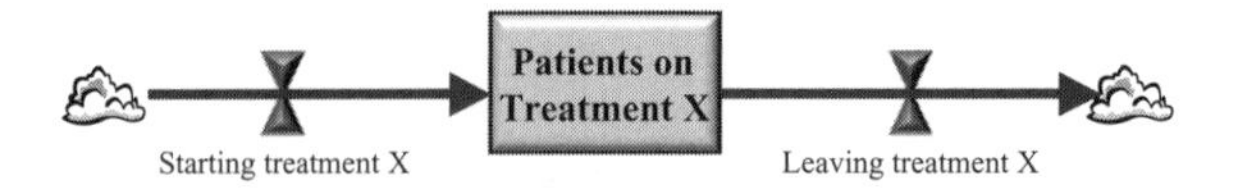

Figure 6 Example dynamic model demonstrating impact of duration of therapy inputs.

the model related to the parameter that controls the therapy duration for Patients on Treatment X and even perhaps the rate of patients starting Treatment X. For the sake of illustration, suppose that the mean duration of therapy is chosen as the best approximation of "average" observed patient behavior*. An exponential random distribution, with the appropriate arguments based on analysis of the historical patient data, can then be incorporated for that input parameter. (Current simulation engines offer a wide variety of random functions specifically designed for this type of analysis.) Similarly, a *uniform* or *normal* distribution might be used to introduce variability into the parameter governing the flow of patients starting Treatment X.

Multiple runs of the simulation must then be performed since there will no longer be a single, unique set of inputs to the model structure. One way to perform this analysis is for each and every patient to take on exactly the same therapy duration in a single simulation run, the value of which is pulled from the appropriate exponential distribution. A more robust analysis involves assigning various cohorts of patient groups a unique therapy duration—again, derived from the appropriate random distribution—at various time points within a single simulation run. This approach has the advantage of being able to assess outcome variability within a single run of the probabilistic model, not simply from comparing multiple simulation runs. Most importantly, patients in a traditional Dynamic Modeling approach are treated on the group level (which may be the result of disaggregation), and not as individuals. Chapter 9 addresses how a dynamic model can be "individualized" to treat each patient as its own distinct entity.

Once multiple simulations are performed, the data can be collected and the results analyzed. As shown in Figure 7, the probabilistic model will now report a range of possible outputs for Patients on Treatment X, giving the marketing team some assessment of the variability of potential outcomes given uncertainty regarding model inputs. This process allows confidence intervals, most likely outcomes, and even analysis of outlier cases to be rigorously evaluated and ultimately implemented into a Brand Planning strategy.

Increasingly, dynamic models are incorporating probabilistic inputs into defined stock/flow structure in order to give more rigorous assessments of expected

* The appropriate measure of central tendency in a various random distributions (such as the exponential distribution implied by Figure 5) is beyond the scope of this work, but interested readers are encouraged to consult any introduction statistics book for more detail. As a practical matter, we often conduct simulations using first the mean, then the median, and finally the mode to determine if the choice of "average" significantly affects the results generated by a dynamic model.

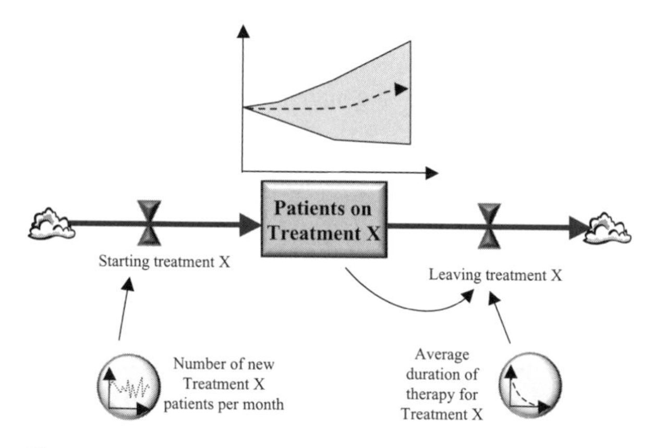

Figure 7 Results of example dynamic model regarding probabilistic inputs and corresponding variability of outputs.

outcomes over time. Data availability, improved computing power, and associated analytic software has made the resulting simulation engines not only faster but also easier to use. As long as these efforts are designed to augment and leverage existing data sources for the purposes of improved decision making (as opposed to simply replicate the complexity of the disease marketplace in a computer environment), integrating uncertainty into the Standard Template framework can often make the associated dynamic model much more robust.

Strategic Implications of Obtaining Valid Patient Flow Probabilities

A central tenet of the Dynamic Modeling methodology is that it is the flows of things in the real world that dictate how systems change over time. In pharmaceutical marketplaces, these flows generally involve patients as they initiate, change, discontinue, or reinitiate treatment. As such, the probabilities of patient flow movement in a disease market are the fundamental driving principles behind the changing marketplace dynamics. Patient Flow probabilities can be used to establish lines of therapy progressions, determine source of volume of patients, and analyze the magnitudes of patient movements that any marketing strategy should be addressing. When these derived metrics have a solid foundation in representative patient-level databases, decision makers can be sure of the robustness of the analysis, helping them make more informed decisions regarding the marketing initiatives that will be the part of a Brand Plan.

The sources and types of data required to populate a Standard Template dynamic model are indication specific, and often vary depending on the goal of the modeling project and the strategic questions it is attempting to answer. This chapter has examined some of these sources and associated data formats required to populate a typical stock/flow structure. Utilizing these types of sources will help ensure that your modeling efforts are both structurally sound *and* backed by solid and defendable data analysis. Examples in subsequent chapters that extend or depart from the Standard Template framework require additional data from a wide variety of sources, and will be discussed in conjunction with those specific examples.

SUMMARY

- Dynamic models require a set of initial conditions that define the status of all the stocks in a system at a single time point.
- Published epidemiology studies, syndicated reports, patient-level databases, and primary market research data are excellent sources to obtain the initial numbers of patients in the stocks of the Standard Template.
- Data derivations for the epidemiology flows associated with the Standard Template were examined in chapters 4 and 5.
- Patient flow treatment dynamics are important drivers of patient and market share.
- Data needed to calculate these dynamic probabilities are best derived from analysis of longitudinal integrated claims databases.
- Patient behavior among treatment options can be conceptualized and analyzed using a Treatment Change Matrix.

7

Specific Models of Patient Flow Dynamics

INTRODUCTION

This chapter will cover the following topics:

- Event-based decision tree
- Medical procedures
- Trial-repeat structures
- Generation of prescription sales

Dynamic Modeling of patient flows can extend beyond the basics of epidemiology, treatment changes, and compliance/persistence calculations. In fact, the methodology can be used to analyze any series of processes/stages through which patients must pass in order to be of strategic importance. For example, in the late 1990s, Novartis conducted a patient flow examination related to its oral antifungal product, Lamisil®. The analysis revealed that patients were reluctant to discuss their condition of toenail infections with their physicians, who in turn often dismissed the problem as a minor cosmetic issue rather than one justifying pharmaceutical intervention. The detailed patient flow analysis suggested the drug's marketing campaign be repositioned to inform consumers about the need and availability of a new treatment for their condition, as well as educate physicians on the concerns and wishes of their patients (1).

This chapter will present three simple examples of dynamic models, which depart from the Standard Template or extensions thereof. These examples are by no means comprehensive, but are included to demonstrate the wide range of issues that can be analyzed using the Dynamic Modeling methodology.

Examples of Patient Flow Modeling Outside the Standard Template

Examples include the following:

- Event-based decision tree
- Medical procedures
- Trial-repeat

EVENT-BASED DECISION TREE

In certain disease areas, the stock and flow structure related to prevalence and incidence is not as relevant from a strategic perspective. In particular, some indications are more **event-driven**; they result from a single and often major physiologic malfunction which requires immediate attention. Examples of event-based or incident-driven diseases for which pharmaceutical intervention must be quickly applied are:

- Heart attacks
- Strokes
- Aneurisms
- Embolisms

For such events, the window for effective treatment application is usually measured in hours, mortality in the critical acute period is often high, and the stages a patient goes through before qualifying for such treatment are often executed in minutes. Clearly, including such short timeframe dynamics in the Standard Template which usually calculates patient flow movements, on a monthly or yearly basis, is impractical. However, Dynamic Modeling technology allows for the choice of timeframe to be matched to the important time metrics of specific indications.

In general, models of event-driven dynamics must include structure to represent the stages and potential dynamic pathways to treatment through which patients might progress. The starting point for such a framework begins with the event itself and tracks patients as they flow through a defined structure, which often has a few standard stages.

Common Stages for Event-Driven Dynamics

Common stages consist of the following:

1. Recognizing the physiologic problem.
2. Deciding on an appropriate course of action.
3. Entering the health care system through an access point.
4. Completing a clinical diagnosis of the problem.

An example of a stock/flow diagram depicting such dynamics is shown in Figure 1.

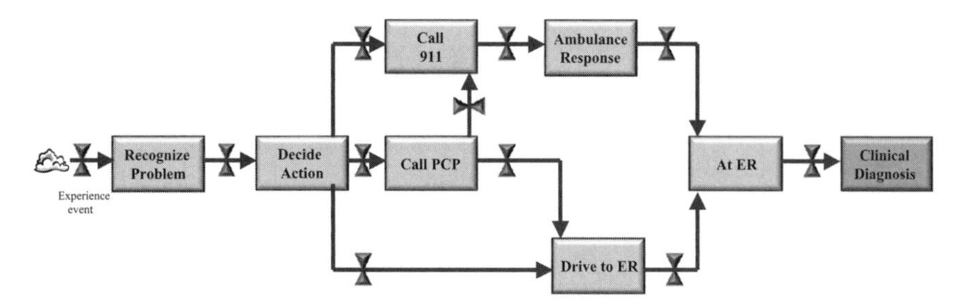

Figure 1 Example for stock/flow representation of patient flow dynamics associated with event-driven indications.

Many readers will note this structure is not entirely dissimilar from a typical decision tree diagram. In fact, this dynamic template is simply a stock/flow representation of that approach, but with the added power of a simulation model. The dynamic framework in this case provides both a visual representation of the process and a computational tool that can be used to test the effectiveness of various intervention strategies. Note that the stocks in this Dynamic Modeling framework represent where patients are at any particular time point, and the flows represent potential pathways as patients move from stage to stage. Often such structures include detailed flow dynamics occurring within the hospital, as patients progress from admission to various waiting stages and areas of the emergency room center. These additions are an easy extension of the basic stock/flow framework depicted in Figure 1.

Data collection for this structure mirrors the expansion of the Undiagnosed Patients structure in chapter 5, but has a much different timeframe that applies to event-driven situations. As in that example, reliable data on the initial values of the stocks/stages in this system is nearly impossible to collect, both in terms of measurement error and the short timeframe associated with such event-driven processes. Instead, the average time patients spend in each stock/stage needs to be collected and from that data, a distribution of patients across the defined phases can be derived from the simulation model. For stocks with more than one possible outflow, the allocation of patients as they exit the stage must also be determined. Figure 2 shows the stock/flow representation of the process, along with associated sample data.*

As Figure 2 shows, the large amount of time spent, on average, in the Recognize Problem stage is an obvious intervention point,† but other potential strategic levers are less clear. Looking at this stock/flow structure and associated

* For simplicity, this structure does not show or include death rates at each stage, but inclusion of mortality is an easy extension.

† Readers may be familiar with television or print advertisements emphasizing the warning signs of strokes, part of a nation-wide awareness campaign on the part of the National Stroke Association (7) to decrease the time patients spend in the "Recognize Problem" stage and urging a quick response by immediately calling 911.

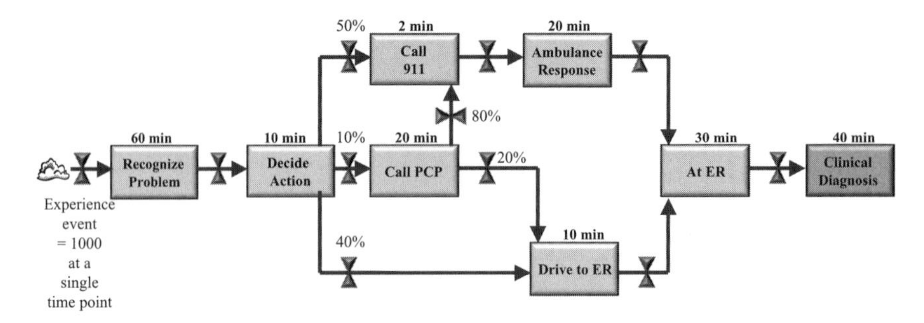

Figure 2 Example event-driven patient flow model dynamics with sample data.

data can give some insight into the relevant patient dynamics, but determining how long it takes, on average, for patients to receive a clinical diagnosis or the distribution of those timing metrics is difficult to determine from mental simulation or static calculations alone. Constructing this simulation model allows not only these metrics to be calculated, but also provides a means for various strategic options to be rigorously tested and evaluated.

One interesting output from such a model is a **dynamic diagnostic curve**, which shows the distribution of time to diagnosis for a cohort group of patients. This plot shows how a group of 1000 people who experienced an event-based problem at exactly the same time would differ in the subsequent time at which the Clinical Diagnosis was completed. The example model shown in Figure 2 produces the dynamic diagnostic curve as shown in Figure 3.

Figure 3 shows the time at which diagnosis is complete on the X-axis, measured in minutes since the event occurred. The Y-axis measures the number

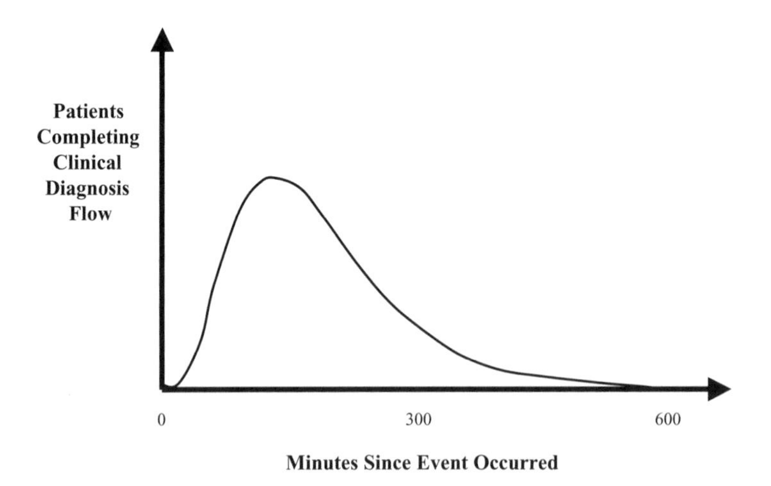

Figure 3 Example event-driven dynamic diagnostic curve.

Table 1 Temporal Band Outputs for Example Event-Based Dynamic Model

Time duration for completing clinical diagnosis (min)	Patients (%)	Cumulative percent of patients
0–60	6	6
61–120	30	36
121–180	31	67
181–240	18	85
241–300	9	94
301–600	6	100

Source: Analysis of example dynamic model results.

of patients (out of 1000) completing the diagnosis in each given time listed on the X-axis.* The time at which diagnosis peaks is about 120 minutes postevent, meaning the greatest flow of patients complete the diagnosis two hours after the event has occurred. This curve can be analyzed in a number of different ways, as shown in Tables 1 and 2.

For many indications, these model outputs and analyses can be compared and/or calibrated to published studies on the aggregate patient arrival times for event-driven occurrences. On a broader scale, the simulation model can be converted to actual patient numbers by applying data on the number of events of a particular type occurring in a given marketplace over a specified interval. For example, the American Heart Association estimates that over 1.2 million heart attacks occur each year in the United States (2). Integrating this type of statistic with the dynamic simulation model would provide the absolute numbers of patients flowing through the process, as well as quantify the effect of possible intervention

Table 2 Quartile Band Outputs for Example Event-Based Dynamic Model

Cumulative percent of patients completing clinical diagnosis	Within number of minutes postevent
25	100
50	145
75	200
100	600

Source: Analysis of example dynamic model results.

* Readers with nostalgia for their college calculus class will recall that the area under the dynamic diagnostic curve represents all 1,000 patients tracked in this analysis.

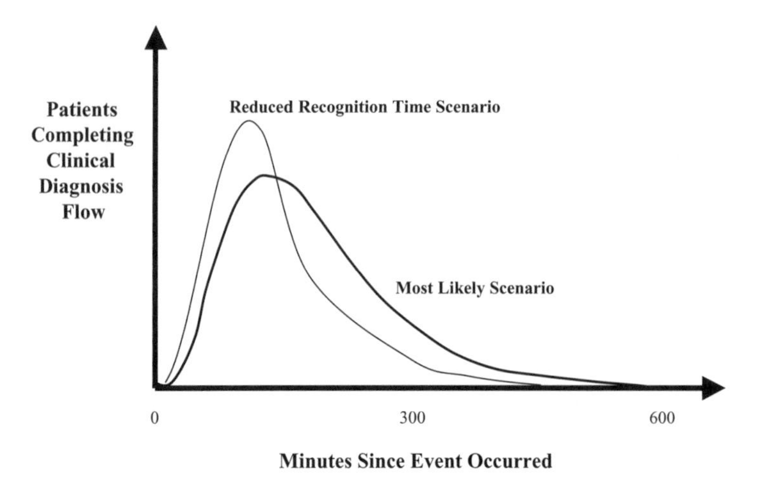

Figure 4 Example event-driven dynamic diagnostic curve—Most Likely Scenario vs. reduced recognition time scenario.

points in terms of market sizing. These numbers are especially effective when converted into financial metrics based on cost-effective calculations.

With a simulation model and Most Likely Scenario (see chap. 11) established, we can now show the impact of parameter changes on the entire dynamic diagnostic curve. Figure 4 shows the effect of reducing the average time in the Recognize Problem stage from 60 to 30 minutes on the Most Likely Scenario. Note how this change shifts the entire curve to the left, with the peak of diagnosis now at about 100 minutes postevent. As would be expected, the cumulative number of patients completing diagnosis at any time point is now greater than in the Most Likely Scenario.

A similar set of comparisons could be completed on changing outflow allocations, such as the percentage of patients driving to the emergency room (ER) versus calling 911 versus calling a primary care physician (PCP), for example. Examining the effect of changing input parameters on the dynamic diagnostic curve begins to show the strength of various strategic intervention points in the system, and evaluating the sensitivity of overall patient dynamics to such changes is a powerful end-point of such a simulation model.

The analysis of the effects on the dynamic diagnostic curve can be extended for parameter testing based on a chosen metric. For example, suppose we are looking at an indication in which effective treatment must be administered within three hours from the time the event occurred. Such "treatment windows" are not unusual in event-driven indications. For example, Genentech's Activase®, also commonly referred to as tissue plasminogen activator, is a thrombolytic agent which must be administered within three hours of the initiation of an ischemic stroke. The simulation model can become a scenario analysis tool to test the effect of changing various model parameters on the number of patients who finish

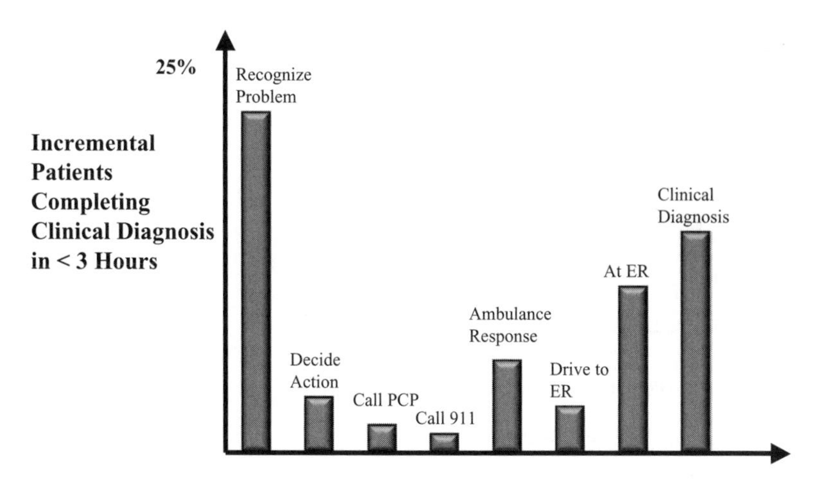

Figure 5 Example sensitivity results of event-driven patient flow model on patients completing Clinical Diagnosis in less than three hours.

Clinical Diagnosis within that 180-minute window. A complete sensitivity analysis changes all parameters, one at a time, by reasonable amounts to test their impact on the number of patients completing the Clinical Diagnosis stage within that critical time window. Figure 5 shows the result of sensitivity analysis on the time spent in each stage, with the average time constant for each phase reduced by 50%.

Some human intelligence must be applied when analyzing the results of this approach, however. For example, it is unlikely that any strategic marketing campaign to reduce ambulance response time ("Drive faster!?!??") will be well received or highly effective, regardless of its theoretical impact on model results. Additionally, reasonability checks must be established when analyzing such sensitivity results, as some input parameters might be more changeable than others. Reducing the time spent at the ER by 50% might be extraordinarily difficult, for example, but reducing the average Clinical Diagnosis time by the same amount might be achievable through medical education and/or new diagnostic tools. Straight sensitivity analysis results begin to give insights into strategic implications, but must be interpreted and sanity checked. Our clients often use a **Strategic Sensitivity Grid** to help with this interpretation process, as shown in Figure 6.

The Strategic Sensitivity Grid takes model sensitivity results such as those in Figure 5 and evaluates them on dimensions of feasibility versus effectiveness. The Potential to Change axis is often measured in terms of expected marketing spend needed to change a parameter by a given amount. The Impact/Sensitivity On Strategic Metric axis then represents the effect of that parameter change on a chosen model metric. Variables with a high potential to change (i.e., low cost) and a high impact on the system (i.e., most sensitive parameters) fall into the zone of the highest strategic leverage in the upper right corner of the grid. Our client

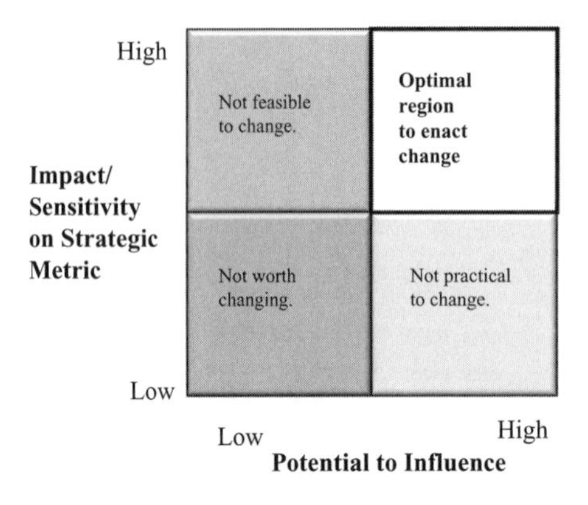

Figure 6 Conceptual Strategic Sensitivity Grid.

teams have found that the Strategic Sensitivity Grid provides a nice conceptual framework for evaluating dynamic model outputs and associated sensitivity results against the real-world backdrop of limited resources.

The sensitivity analysis shown to this point involved changing individual parameters, one at a time and in isolation, but combinations of highly leveraged and achievable intervention points can be examined as well. The dynamic simulation model provides a means to test such combinations, which in some cases can show synergies between strategic options. In other words, the effect of combining Strategy A and Strategy B might be different than simply adding the numbers from those individual sensitivity results. Dynamic processes have behaviors that are often difficult to intuit without the help of simulation, so comprehensive testing and evaluation of model results is essential to successful and effective strategy formation.

Dynamic models can also be used after the fact to examine how strategic interventions affected a marketplace. Returning to our continuing example, suppose a marketing campaign was implemented to reduce the Recognition Time of an event-driven problem in a given geographic region. A data collection process at all local hospitals would be able to determine the degree to which the dynamic diagnostic curve shifted as a result of that campaign. Such results could then be calibrated to the dynamic model, determining the degree to which recognition time was actually reduced. These metrics tracking exercises are another way to leverage the use of the simulation tool, and are frequently used to analyze the economic effectiveness of such intervention strategies.

By combining a traditional decision tree structure with the power and flexibility of Dynamic Modeling simulation allows for insightful analysis and a unique tool for strategy development in event-based indications. This approach can easily be adapted and extended to address-detailed aspects of event-driven patient dynamics such as in-hospital patient flows, rural versus urban settings, and diagnosis/pharmaceutical intervention by ambulance personnel. In event-driven marketplaces, the stock/flow model structure provides a framework for conceptualizing key processes, as well as a simulation tool to provide analysis and test effective intervention strategies.

MEDICAL PROCEDURES

The Standard Template stock/flow structure related to prevalence and incidence does not fully capture the complexities of all marketplaces. Medical procedures, for example, are one-time events taking place in a health care setting such as a hospital or surgery center. While not as extensive as the markets for chronic indications, the number of patients experiencing such procedures is large and growing due to aging of the population and advances in medical techniques. In 2005, for example, over 699,000 open-heart procedures were performed in the United States (3). Examples of common medical procedures are:

- Surgeries
- Transplants
- Joint replacements

While data on the occurrence of the procedures themselves is often relatively easy to obtain, the number of patients surviving the initial intervention is notoriously difficult to estimate, as many survivors have limited contact with the health care system after a successful procedure. Static approaches to such analysis often involve complex tracking of cohort groups and estimations of resulting prevalence metrics. Incorporating epidemiology metrics that might change over time such as survival rates, subsequent procedure probabilities, and/or changing patient categorizations (acute vs. maintenance) becomes increasingly difficult with a spreadsheet methodology. The inherent time-based nature of the Dynamic Modeling approach, however, is well suited for analysis of medical procedures marketplaces, and provides a useful framework to conceptualize and analyze such situations.

Medical procedures are typically performed in a hospital setting, with post-procedure patients remaining for some period of observation time. Mortality during this period is generally low but not discountable, and most patients are discharged and actively tracked for some period of time in a critical, postoperative acute phase. Patients surviving the acute stage go on to a maintenance phase, where they remain until death. A sample of the aggregate stock/flow structure capturing these important medical procedure dynamics is shown in Figure 7.

The structure shown in Figure 7 can be expanded to include some detailed aspects associated with medical procedures—depending on the needs of the working team and the markets they are investigating. In our experience with clients, we have extended the basic stock/flow framework as follows:

- For pharmaceutical products that might be used in conjunction with surgeries, the stocks of patients at each stage can be disaggregated to account for

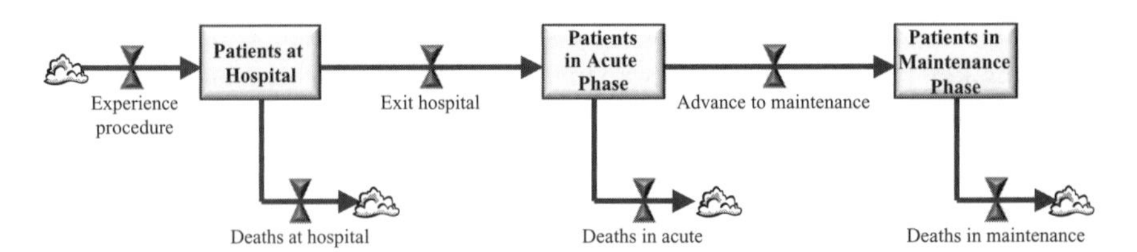

Figure 7 Example aggregate stock/flow structure for medical procedure patient flow dynamics.

prescription therapy usage and treatment change dynamics (see chap. 4). For example, specialized blood thinning medications are often used in patients undergoing knee reconstructions.

- For transplant markets, the dynamic model can be expanded to include dynamics around transplant rejections, donor programs, waiting lists, and/or associated treatment regimens. For example, the United Network for Organ Sharing estimates there were approximately 2620 people on the waiting list for a heart transplant in 2008, but only approximately 2200 such operations were performed (4).
- For joint replacements, the stock/flow structure can be extended to show the dynamics surrounding procedure failure rates at the time of the replacement, average lifetime of a joint replacement, and subsequent reprocedure rates. For example, industry estimates are that less than 10% of hip replacement patients require revision surgery at some point during their lifetime (5).

Case Study: The Dynamic Modeling Approach to Medical Procedures

For illustrative purposes, suppose a medical devices firm is developing a new hip replacement prosthesis using recent advances in materials sciences. The device will not be commercially available until 2015, yet the marketing team is being asked by upper management to provide an evaluation of market potential as well as a preliminary strategic plan for market penetration. As a first step, the working team pieced together some historical data from sources such as the U.S. Census Bureau, the American Academy of Orthopaedic Surgery, and National Hospital Discharge Survey to estimate the number of hip replacements taking place in the United States over the last 15 years, as shown in Table 3.

The data in Table 3 could certainly be extrapolated to a forecasted projection of the number of hip replacements occurring in the year 2011 and beyond. But a simple forecast might mask an important strategic dynamic; *patients whose existing hip replacement has worn out and now requires a replacement.* The conventional industry wisdom is that only 10% of hip replacement surgeries are performed on patients to replace their original prosthesis, and such a small flow of patients may not be interesting from a strategic perspective. But depending on the mortality rates for hip replacement patients, changing ages at which patients first experience such procedures (particularly the age 75–84 demographic), and

Table 3 Number of Hip Replacements in the United States

Year	Hip replacement surgeries
1990	119,000
1991	117,000
1992	127,000
1993	125,000
1994	124,000
1995	134,000
1996	139,000
1997	144,000
1998	160,000
1999	168,000
2000	152,000
2001	165,000
2002	170,000
2003	177,000
2004	185,000
2005	193,000
2006	199,000
2007	208,000
2008	217,000
2009	225,000
2010	235,000

Source: National Hospital Discharge Survey (NHDS) (1990–2004), extrapolated through 2010 using population data from the U.S. Census Bureau and estimates from the American Academy of Orthopaedic Surgery.

expected functional life of the devices themselves, such reprocedure dynamics could represent an increasing source of patient volume in the future. A strictly numeric forecast based on extrapolation of historical data, while potentially useful from a market sizing perspective, would not explicitly capture this key dynamic in the hip replacement marketplace.

In order to understand and quantify the strategic implications of such an evolving indication, an operational analysis that goes beyond simple statistical forecasting may be required. In fact, from a strategy development standpoint the shortcomings of a simple data extrapolation suggest the need for the following:

- *An integrated model to check the internal validity of parameter assumptions.* Does the conventional wisdom that 90% of hip replacements are performed on first-time patients make sense relative to the data on the number of procedures taking place now and in the recent past? How robust is this estimate of the

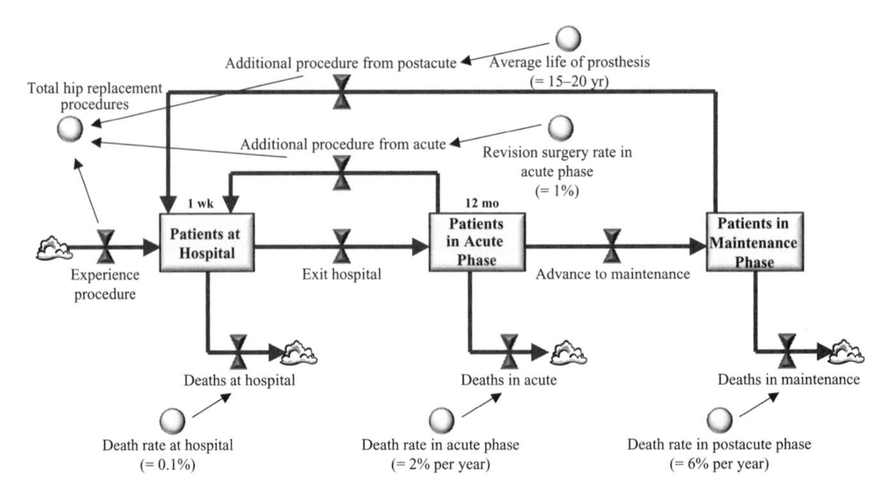

Figure 8 Example expanded stock/flow structure for hip replacement patient flow dynamics and associated sample input data.

revision surgery rate, given ranges of uncertainty around other data assumptions? Figure 8 shows an extended stock/flow framework and sample input data that will allow such questions to be addressed and ultimately answered.

- *The ability to incorporate input parameters that might change over time.* How do falling mortality metrics (*death rate in postacute phase*) and increasing success of existing hip replacement devices (*average life of prosthesis*) change the estimated number of patients requiring an *additional procedure from postacute*? Do trends in the average age at which patients experience a hip transplant affect the expected dynamics of the marketplace as well?

Any of the input auxiliary variables in Figure 8 can be established as scenarios based on recent trends in their magnitude. The inflow to the Dynamic Modeling structure of patients who *experience first procedure* is an obvious candidate for this approach, given historical trends in total numbers of hip replacement surgeries. This variable could also be operationalized by relating it to population dynamics as well as demographic shifts related to aging of the U.S. population. Other auxiliary variables such as the *average life of prosthesis* metric, for example, can be established as a time-variant input as shown in Figure 9.

The dynamic model shown in Figure 8 can be simulated to show expected evolution of the hip replacement marketplace, given established operational structure and associated input data. Under one particular set of assumptions, the simulation outputs for 2015 are shown in Figure 10.

Note that the variable *total hip replacement procedures* in Figure 9 ties back to Table 3 and can be calibrated to extrapolations of historical data on the number of surgeries in this marketplace. Of greater importance from a strategic perspective, however, is the flow of patients experiencing an *additional procedure*

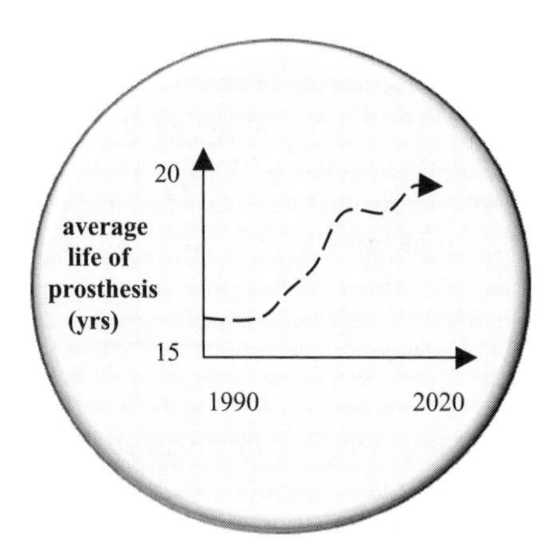

Figure 9 Example of time-based scenario for *average life of prosthesis* variable.

from postacute. The dynamic model simulation results suggest this reprocedure component of the marketplace may represent almost 20% of all hip replacements in 2015—almost double what current estimates suggest. The marketing approach of our example medical devices firm might be influenced by such a finding, as efforts to reach patients experiencing a revision surgery and/or the doctors that treat them might be completely different than campaigns directed at first-time patients. Dynamic model results of this type often help inform strategic direction in terms of product segmenting, positioning, and even clinical trial design. By operationally defining the processes of patient flows within the hip replacement marketplace, a working team often identifies important insights regarding possible market evolution.

Operational dynamic models, which focus on the fundamental drivers of marketplace change, provide a level of insight and strategic direction that might be

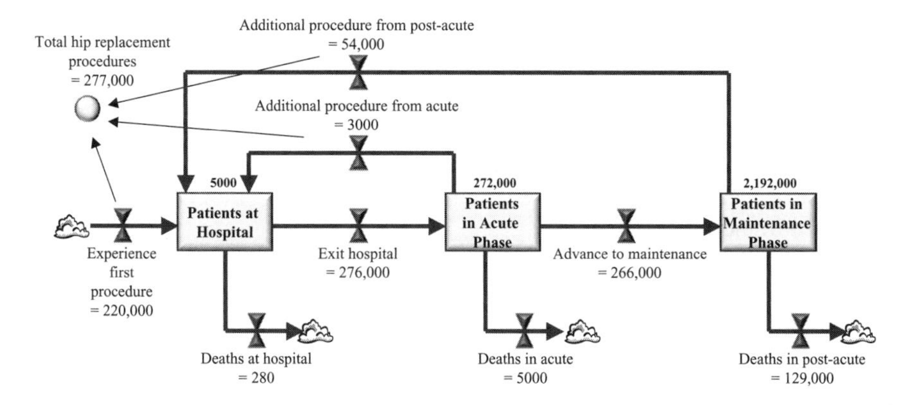

Figure 10 Example hip replacement dynamic model and associated outputs (2015).

hard to derive from a simple forecasting exercise. In our client experience, patient flow models of medical procedures such as hip replacements prove to be a useful framework to check the validity of various input data and incorporate changing marketplace metrics which might affect how a particular indication market might evolve over time. When these insights are translated into actionable strategies, the Dynamic Modeling methodology becomes an integral part of a firm's strategy development process.

TRIAL-REPEAT MODELS

Many of the example models discussed thus far deal with existing marketplaces; indications where current and future therapies compete for patients and share. However, there are some diseases/conditions for which prescription therapy has not yet been developed, and hence marketplace data and analogies are nonexistent. A number of marketing models have been developed to address these types of situations (6), but these approaches are usually applied to consumer products and are often intimidating to those without strong technical backgrounds. Converting such models into a Dynamic Modeling framework provides both a more visually intuitive picture of the marketplace and a useful tool for strategic analysis and short-term forecasting. One such translation involves a trial-repeat framework for a newly released prescription product.

Case Study: Trial-Repeat Frameworks in Pharmaceutical Marketplaces

Imagine a market for a cosmetic condition traditionally treated only with non-prescription therapies, or in which pharmaceutical treatment dealt only with the symptoms, not the underlying cause, of the condition.* In this type of "untapped" market, issues surrounding the diagnostic process or switching to other products are of lower importance. Of greater interest are items such as marketing-induced awareness of the product, response rates, delays in seeing a doctor about a prescription, likelihood of renewal, etc.

Each of these elements might be of strategic importance in developing a Brand Plan for an NCE, both as a means to quantify the potential of the compound in the marketplace and also to develop effective strategies for marketing initiatives. In the absence of historical analogs, such components must be explicitly addressed and accounted for in order to have an informed basis for brand positioning.

In such cases, a stock/flow structure as depicted in Figure 11 often provides a useful framework for analysis.

Initially, the entire Prevalent Population is in the Unaware of Product stock, as the product is not yet on the market. Marketing efforts push patients into the Aware Population category, some of whom are categorized as having No Intent to Purchase because the ad campaign was not sufficient to generate an interest in

* This hypothetical example is presented for illustration purposes only, and is not intended to describe or represent the Lamisil story from the introduction to this chapter.

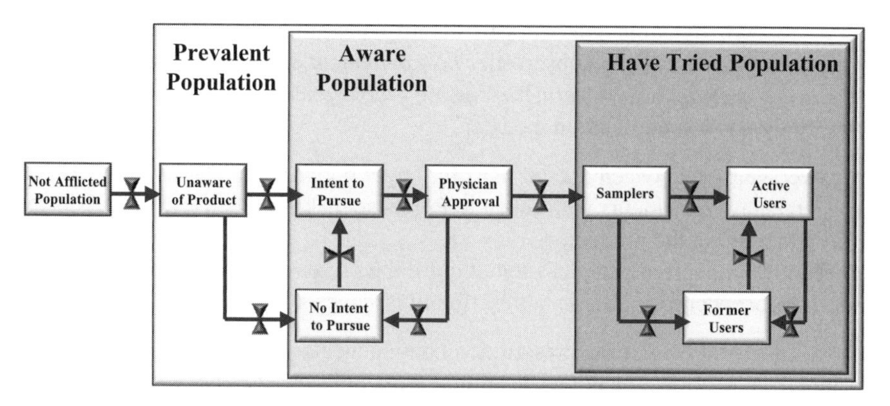

Figure 11 Example stock/flow framework for trial-repeat process.

obtaining the product in question. Other potential patients are initially classed as having an Intent to Purchase because their exposure to marketing has generated a purchase intent.* In prescription marketplaces, the patient must now get Physician Approval for the prescription, either at a specially scheduled or regular doctor visit. Depending on product profile, willingness to prescribe, and other physician-related variables, a prescription for the product may or may not be issued. Falloff between writing and filling of a prescription is often high in these types of indications, but patients who do ultimately follow through on their issued prescription enter as Samplers for a period of time usually defined as the duration of the initial prescription. Those patients not receiving Physician Approval return back to the No Intent to Purchase category.

The dynamic model structure to this point addresses the initiation of the *trial* of a product. The remaining portions deal with the *repeat* process. Samplers can drop off and become Former Users, or can decide to continue to use the product and become Active Users. Active Users are subject to subsequent drop off as well, but perhaps at a reduced rate from Samplers. Former Users can reinitiate treatment and re-enter the Active Users stock at some later date, although the metrics that govern this process are impossible to determine before product launch.

The data needed to populate such a dynamic structure usually come from primary market research, with secondary studies of comparative products filling in the blanks. First, data regarding the percentage of the focus population that will ultimately be issued a prescription for the product will need to be collected, such as:

- Awareness generation curve
- Target population response to advertising
- Physician metrics regarding likelihood to prescribe

* This description relates to the *initial* response to advertising, but some potential patients could initially be resistant to the product but decide later on to pursue it. This dynamic is captured by the flow from No Intent to Pursue to the Intent to Pursue stock.

Second, dynamic metrics will identify the time frame over which the patients' responses will be observed. *This component of the model becomes particularly important if short-term forecasting is an explicit goal of the model.* These type of dynamic metrics often include:

- Average time between doctor visits for target population.
- Percentage of patients with Intent to Purchase making specific doctor visits to inquire about the product.
- Time delay before patients attempt to fill the prescription.
- Drop-off occurring between prescription issuance and fulfillment.*

Third, data regarding prescription continuance and reinitiation may be estimated from historical analogies in similar marketplaces, such as:

- Expected treatment discontinuation rates, for either Samplers or Active Users
- Reinitiation of treatment by Former Users

Parameter estimates to populate dynamic trial-repeat models of emerging marketplaces are at best tricky, as by definition these markets do not have any historical analogs or existing data on patient behavior. As such, input estimates are often the result of team judgment rather than hard data collection efforts. Simulation models can be used to test the importance of various inputs to overall market performance through sensitivity analysis, as we discussed earlier in this chapter.† This type of sensitivity analysis can quantify the impact of model input parameters, effectively prioritizing the need for better data and focusing the resulting data collection efforts on the most sensitive input parameters. Subsequent primary market research can then be designed so that data on important metrics is collected, analyzed, and integrated into the evolving dynamic simulation model. In this fashion, the Dynamic Modeling methodology becomes an evolving framework not only for a priori quantification, but also for data prioritization and strategy development.

The revenue generation algorithm in a dynamic model focusing on such short-term forecasting, and in which monthly precision may be important, takes on a different flavor than has been shown thus far. In the real world, pharmaceutical revenues are generated when patients fill and pay for prescriptions, either out of pocket or in combination with insurance reimbursement. Subsequent revenues result when patients refill their prescriptions for a particular medication. As shown in chapter 4, many dynamic models aggregate this process to depict revenue generation as Currently Treated Patients multiplied by an average monthly price of therapy and then a true compliance rate. In many trial-repeat contexts, a more precise formulation may be needed. First, notice that from an operational standpoint,

* This concept is often referred to in pharmaceutical circles as the degree of adherence–the likelihood of filling an initial prescription.

† In addition, chapter 14 introduces some expanded techniques for deriving parameter estimates applying to latent marketplaces.

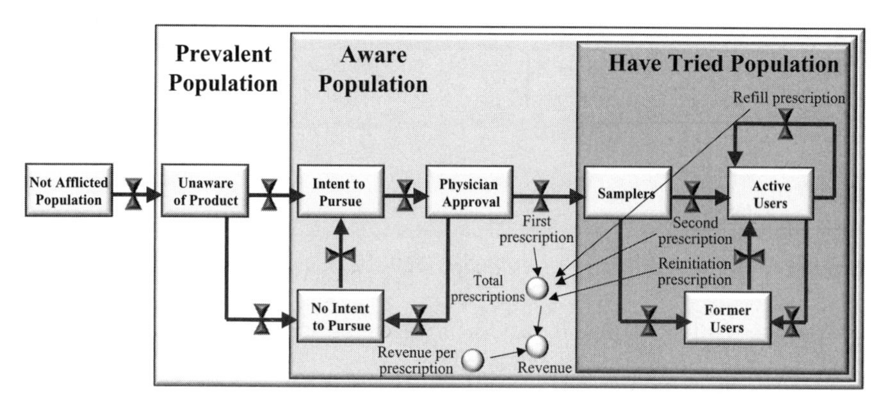

Figure 12 Dynamic model trial-repeat structure with prescription-based revenue generation.

revenue is generated from the flow of patients filling prescriptions, rather than the stocks of those patients themselves, as shown in Figure 12.

Each of the labeled flows in Figure 12 denotes a place where revenue is generated. The flow into the Samplers stock occurs when the *first prescription* is issued. The subsequent flow into Active Users depicts the *second prescription.* Thereafter, Active Users may experience *prescription refills*, as shown with a reflective flow that keeps patients in the stock but shows explicitly an important patient dynamic. After discontinuing treatment, Former Users may begin treatment again and receive a *reinitiation prescription.* In this case, the formula for revenue generation is very simple. The total of these prescription flow represents the *total prescriptions* written in a given time period (usually a month), which when multiplied by the *revenue per prescription* will result in *revenue* generated per month.

The only abstract part of prescription-based revenue generation revolves around categorizing the *prescription refills* flow. In general, pharmaceutical products are often prescribed with a 1 to 3 month prescription that can be refilled up to six times. Refills rarely occur precisely as prescribed, however. As described in chapter 4, the issue of compliance takes its toll on revenue generation in the pharmaceutical industry. A patient with 50% compliance, for example, would only refill his script every four months instead of every two months as prescribed. Dynamic Modeling software can integrate this concept by combining a compliance metric with a time-based marker specific to when a patient first received a prescription. Longitudinal patient tracking of this type provides the basis to for a dynamic model to determine when future refill prescriptions should be realized.

For example, suppose the dynamic model calculates 100 patients will receive their *second prescription* and become Active Users on June 1st of a particular year and that in this market 60 day prescriptions are standard. Each of these 100 should, in fact, be getting their *prescription refills* sometime in early August. However, if the average compliance metric is expected to be only 50%, these patients will not

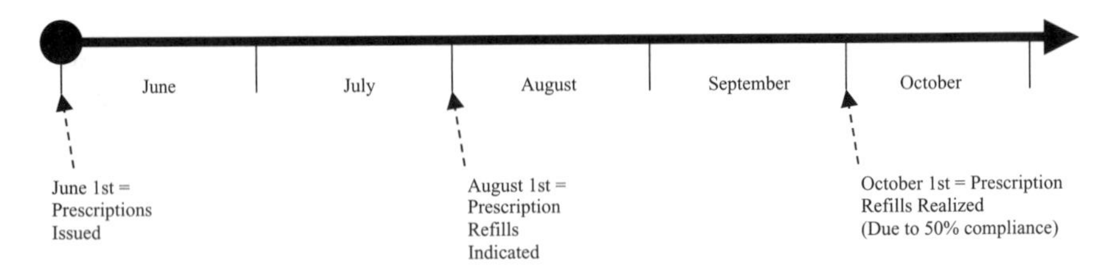

Figure 13 Timeline diagram of prescription refills assuming 50% compliance.

be ready to refill in August. In fact, their current prescription will last them until the first part of October, as shown in Figure 13.

A dynamic model accounts for variable prescription refills by time-stamping each cohort group of patients as they receive prescriptions. Then the simulation software can longitudinally track patients to determine when they will actually need refills based on compliance metric parameters.* Depending on the degree of time-frame accuracy required, such models can be set up to calculate monthly, weekly, or even daily market metrics, including revenue generation.

Similar to the previous examples, this dynamic model structure and resulting simulation can be used to test sensitivity of various parameters, both for the purposes of verifying short-term forecasting and for determining points of high leverage. For example, a series of simulations might show the following possibilities for the generation of scripts over a 4-year time horizon, as shown in Figure 14.

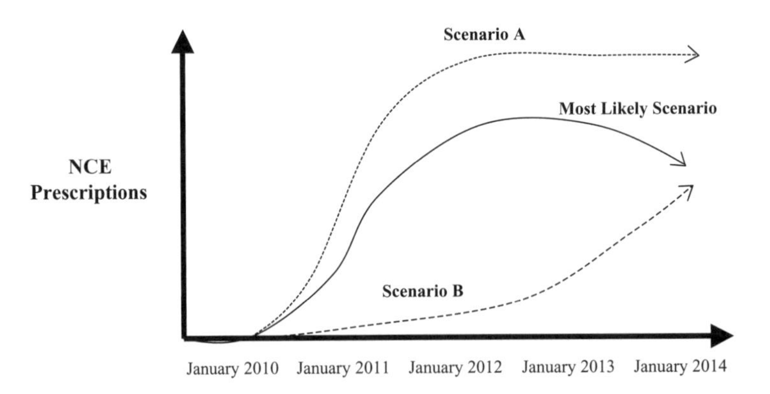

Figure 14 Example simulation outputs for dynamic trial-repeat model under alternative scenario assumptions.

* An easy extension includes a distribution around such metrics, so that the average compliance is 50%, but could be as high as 95% for some patients, as low as 30% for others, for example. Each individual patient would then have his or her own prescription and refill schedule rather than being lumped into the collective aggregate. See chapters 6, 14, and 16 for details on how to include distributions related to model input parameters.

With a revenue calculation attached to prescription generation, Net Present Values (NPVs) of each of these scenarios can be determined, given an appropriate discount rate. Such NPV calculations provide a much clearer picture of the size of the opportunity in a given marketplace, and also a sense of the range of possible financial outcomes. These types of NPV scenario analyses can be extended to include Monte Carlo simulations (see chaps. 14 and 16) to more fully explore the possibilities of dynamic market evolution. The resulting simulation tool can also be used to assist in strategy development by establishing a cause-and-effect model linking actions to expected outcomes. As we have seen in previous chapters, the use of dynamic models as a "virtual testing ground" is one of the most powerful aspects of this methodology.

SUMMARY

- Patient flow dynamics extend beyond those captured in the Standard Template or extensions thereof, and can easily be analyzed using the Dynamic Modeling approach.
- Event-driven indications can be addressed by establishing pathways to treatment for afflicted patients, with dynamics playing out in minutes or hours instead of months or years.
- Medical procedures can be analyzed using stock/flow structure to quantify the dynamics associated with postoperative acute phases, maintenance stages, and reprocedure rates.
- Patient flow models of medical procedures can help estimate the number of patients living in the acute and postacute phases of their recovery, allowing for better evaluations of marketplace sizing in the face of changing market conditions.
- Trial-repeat models can be established to investigate the behavior of patients in latent markets with no existing competitive products and/or useful analogs. Chapter 13 has more specifics on analytic techniques to address such situations.
- Strategic Sensitivity Grids help put the results of sensitivity analysis in a usable context and provide a visual means to assess the tradeoffs between parameter impact versus feasibility.

REFERENCES

1. Simon F, Kotler P. Building Global Biobrands: Taking Biotechnology to Market. New York, NY: Simon & Schuster, 2003.
2. American Heart Association. Heart Disease and Stroke Statistics—Final 2004 data. Dallas, TX: American Heart Association, 2008.
3. American Heart Association. 2008 Update. http://www.americanheart.org.

4. United Network for Organ Sharing. http://www.unos.org; The Organ Procurement and Transplant Network. http://www.optn.org.
5. American Academy of Orthopaedic Surgery. http://www.aaos.org.
6. Lilien G, Kotler P, Moorthy KS. Marketing Models. Englewood Cliffs, NJ: Prentice-Hill, 1992.
7. American Stroke Association. http://www.strokeassociation.org.

8

Dynamic Models of Doctor Adoption of Newly Released Pharmaceuticals

INTRODUCTION

This chapter will cover the following topics:

- Product adoption frameworks/Bass Diffusion model
- Stock/flow structure for Doctor Adoption
- Incorporating physician awareness, trial, and usage (ATU)
- Factors contributing to adoption (internal vs. external)
- Quantifying the effects of physician marketing strategies
- Calibrating dynamic model of Doctor Adoption to historical data

The previous chapters have focused mostly on Patient Flow dynamics, but prescribing physicians clearly play in important role in pharmaceutical marketplace behavior. This fact is not lost on pharmaceutical firms, whose spending on marketing to doctors continues to grow. A recent study concluded that the U.S. pharmaceutical companies spent over $50 billion on promotional activities (much of it directly targeting doctors) in 2004—an astonishing $61,000 per practicing physician (1). These marketing efforts have been under the scrutiny of a number of industry experts to analyze the effectiveness of various types of promotions. Often such analyses take the form of a return on investment (ROI) calculation which correlates promotional spending with resulting revenues, sometimes with a delay (2). The methodology behind these types of calculations may vary, but the common mental model implicit in such analyses is shown in Figure 1.

Such analyses are sometimes useful, but are incomplete as they fail to address of the underlying dynamics which drive the behavior of prescribing physicians in a marketplace. Fortunately, the Dynamic Modeling methodology provides a useful operational framework to explicitly capture the structure behind this promotional activity on the part of pharmaceutical firms. This Doctor Adoption framework can

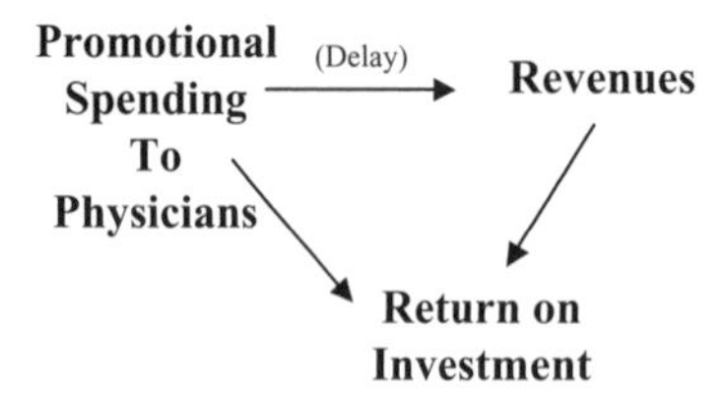

Figure 1 Traditional mental model of physician promotional spending and ROI.

then be combined with the Patient Flow and Treatment Attractiveness sectors to arrive at a fully integrated Standard Template dynamic model.

There is a rich and detailed history of research and data in the area of consumer adoption of newly introduced products. The marketing literature is ripe with studies investigating both traditional adoption/diffusion frameworks and extensions thereof. In essence, these models deal with the rate at which new products are adopted and the factors governing that rate. Perhaps the most popular of these models is the Bass Diffusion model, which despite certain drawbacks has stood the test of time and continues to inform the analysis of new product diffusion in a wide variety of industries. Translating the Bass Diffusion mathematic formulas into a stock/flow framework is shown in Figure 2.*

Figure 2 represents clearly a very simplified version of the product adoption process. Popular extensions of this framework include the option of including people according to their product experience (currently use the product, have never used the product, and previously used the product), the flows associated with those designations, and other potential market segmentations.

In general, Bass Diffusion models further investigate the influencers of adoption, which in broad terms can be grouped according to:

1. *Internal influencers*: word of mouth within the population
2. *External influencers*: the effects of advertising, promotion, marketing, etc.

These influencers are easily included in the dynamic stock/flow structure with the use of auxiliary variables, as shown in Figure 3.

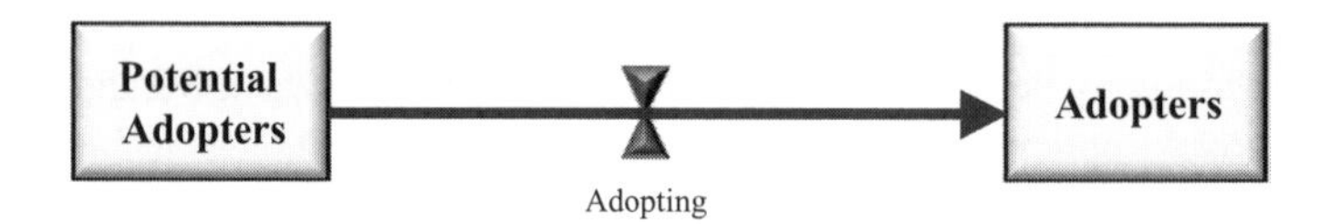

Figure 2 Stock/flow framework of Bass Diffusion model of product adoption.

* This framework is adapted from and explained more extensively by Sterman in his seminal work, *Business Dynamics: Systems Thinking and Modeling for a Complex World* (5).

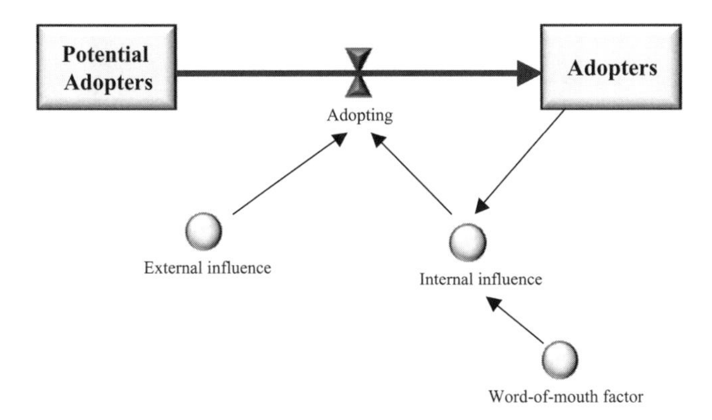

Figure 3 Stock/flow framework for product adoption including aggregate influencers.

In the expanded formulation depicted in Figure 3, the *word-of-mouth factor* variable represents the strength or effect of Adopters on the *adopting* flow. In essence, this variable says, "How many Potential Adopters does each Adopter persuade to purchase the product each time period?" Though not explicitly shown in Figure 3, the *external influence* variable can itself be a function of the levels of marketing spending associated with a product lifecycle. This extension will be included later in this chapter as we expand this continuing example.

The relative strength of the various influencers can be calculated when the aggregate historical data on the actual adoption of products in different market-places is available. This calibration can be done either in closed form or using an optimization algorithm in Dynamic Modeling software packages. The details of this technique are beyond the scope of this work, but at an aggregate level, product diffusion curves can be categorized as shown in Figure 4.

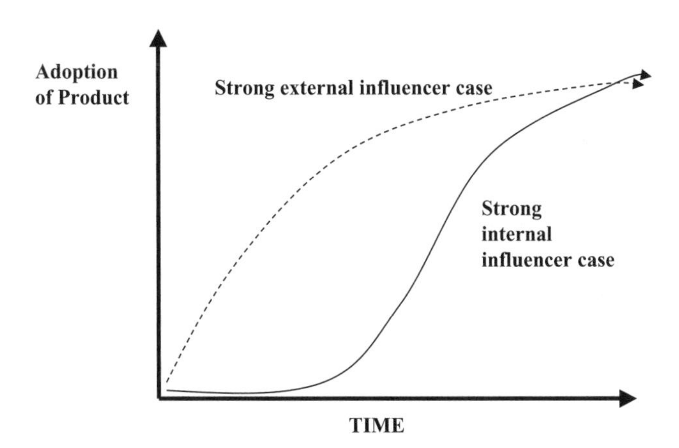

Figure 4 Aggregate product diffusion results.

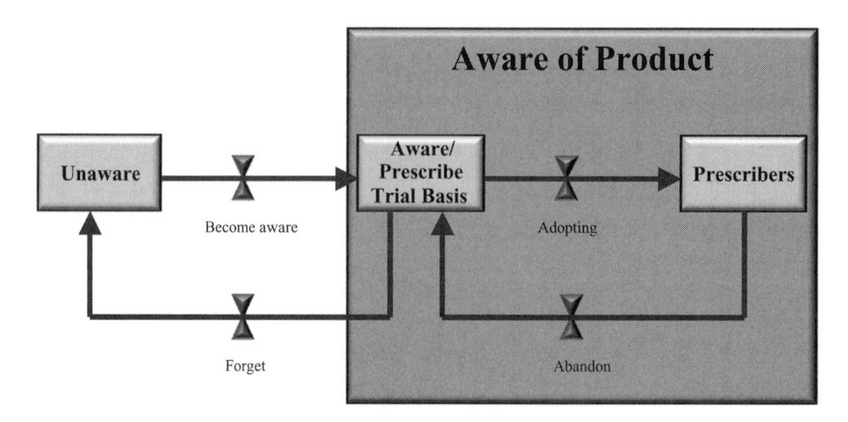

Figure 5 Expanded stock/flow framework for adoption of pharmaceutical products by physicians.

The framework of the Bass Diffusion model provides a powerful set of insights into the factors governing the adoption/diffusion of particular products, the details of which are available in various marketing literature sources. For our purposes, this basic structure can be extended to address the adoption of pharmaceutical treatments by prescribing physicians.

DOCTOR ADOPTION OF PHARMACEUTICAL PRODUCTS

In pharmaceutical markets, newly released treatments must first be accepted by prescribing physicians before they can reach the hands of interested consumers. In effect, doctors are an intermediary and rate-limiting step in the diffusion of products into the marketplace, and therefore require some rigor in analyzing their behavior. We have found that expanding the generic adoption structure of the Bass Diffusion model as shown in Figure 5 provides a powerful framework for strategy analysis.

Some definitions related to Figure 5 and this Dynamic Modeling framework are necessary to ensure proper interpretation.

- The stock/flow structure of Figure 5 is intended to represent the entire universe of doctors who might prescribe a particular drug for a specific indication.*
- Awareness of Product is intended to represent unaided awareness, sometimes known as "top of mind" recognition. We have found that aided or prompted awareness metrics are much too broad to be useful in this methodology.

* Some collective wisdom is required to determining useful inclusion criteria for physicians. For example, markets in which prescriptions are primarily issued by specialists should not include General Practitioners in this framework.

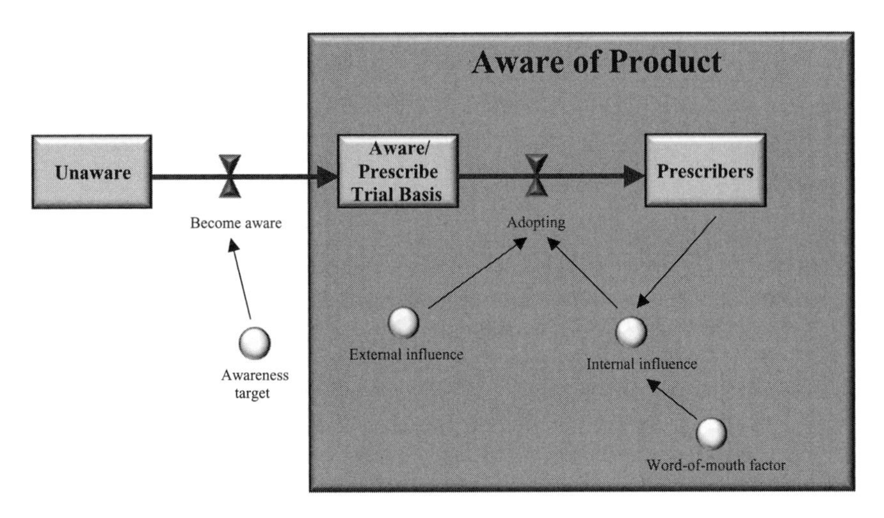

Figure 6 Stock/flow framework for integrating Bass Diffusion model with awareness, trial, and usage of pharmaceuticals by physicians.

- The stock labeled Aware/Prescribe on a Trial Basis represents doctors who know about the product, might occasionally give out a sample or prescribe the drug on an experimental basis, but are not active prescribers of the drug.
- Prescribers are physicians who actively use the drug to treat some, but not all, of their patients suffering from a particular indication. In other words, this designation does not imply that physicians must be exclusive prescribers of a particular drug, only that the drug in question is in their consideration set and thus a viable option when choosing a pharmaceutical treatment for their patients.
- The flows labeled *abandon* and *forget* allow physicians to "retrograde" in their adoption patterns, such that some doctors previously prescribed a certain drug, some doctors used to know about a drug but have since forgotten about it, etc.*

The Doctor Adoption structure shown in Figure 5 can now be expanded to account for historical behavior of pharmaceutical offerings in a particular market. By including the dimensions of product recognition and prescribing behavior, the stock/flow representation replicates the metrics of ATU commonly collected and analyzed by many pharmaceutical companies. Figure 6 is an example of this inclusion.

The *awareness target* variable in Figure 5 is usually a time-based scenario concerning company goals regarding overall product awareness. Operationally, the concept of awareness over time is generated by the rate at which a pharmaceutical

* These flows will be eliminated in subsequent diagrams for the sake of simplicity, but remain in principle unless otherwise noted.

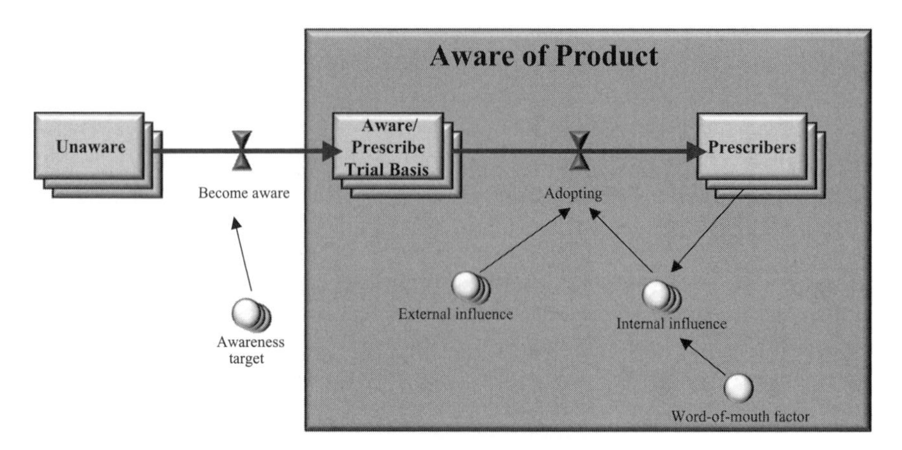

Figure 7 Doctor Adoption framework disaggregated by individual treatment options.

firm can enable physicians to *become aware* minus the rate at which they *forget* about a certain product and return to the Unaware category.* The adoption metrics of *external* and *internal influence* remain from generic Bass Diffusion structure, and can easily be extended to account for various marketing levers at a firm's disposal.

CALIBRATION OF STRUCTURE TO HISTORICAL DATA

The stock/flow structure relating to Doctor Adoption of pharmaceuticals allows for analysis of previously released products in a particular indication. Often such investigations are completed with the aim of forecasting expected adoptions of soon-to-be-released New Chemical Entities (NCEs), and/or determine the factors contributing to successful launches of previously released compounds. Doing so provides marketing teams with a better understanding of this important dynamic in pharmaceutical markets, a tool to forecast expected adoption (based on historical analogies) in a specific indication, and a structure that can be easily integrated into the basic Patient Flow Model framework (see chap. 5). This process allows a Brand Planning team, to more effectively formulate the strategies related to physician marketing as well as understand the impact of those strategies on the disease marketplace.

First, the stock/flow representation in Figure 6 must be disaggregated to include all drug options to be included in the model. Structurally, this is done by introducing a dimension including elements of various treatment options that are of interest in a particular indication, as shown in Figure 7.

* Pharmaceutical firms are increasingly turning to prelaunch awareness as a huge part of their marketing strategy, including large clinical trials to accelerate the adoption of soon-to-be launched compounds.

Table 1 Sample Historical Data on Awareness and Prescriber Status over Five Years in the Marketplace for Example Treatment Options

Years since launch	0	1	2	3	4	5
Awareness						
Treatment 1 (%)	50	80	90	95	98	100
Treatment 2 (%)	0	10	20	25	27	30
Treatment 3 (%)	20	40	80	100	100	100
Prescribers						
Treatment 1 (%)	0	70	90	95	98	100
Treatment 2 (%)	0	1	3	15	25	30
Treatment 3 (%)	0	10	70	97	99	100

For the sake of illustration, we will assume an example indication marketplace that includes three treatment options: 1, 2, and 3.* With the dynamic model structure disaggregated to account for these specific treatment options, we can investigate the past behavior of the marketplace based on historical data.

For an effective retrospective analysis, longitudinal data on both the historical Awareness of the various products as well as the number of Prescribers issuing prescriptions for said treatment options needs to be collected. Typically, such data are organized not in calendar year metrics but rather related to time since launch of each individual drug; often as is shown in Table 1.†

Collecting, displaying, and analyzing Awareness and Prescribing data in percentage terms as shown in Table 1 rather than in absolute numbers of physicians is useful for the following reasons:

- An appropriate universe of prescribing physicians can be chosen, the composition of which could change over time. Some PCPs, for example, eventually begin to prescribe drugs previously only issued by specialists.‡
- Comparison across indications is consistent and easily interpreted. Widespread diseases such as diabetes or hypertension have Prescriber bases numbering in the tens of thousands, while only a handful of doctors prescribe medication for certain rare diseases.
- Model outputs can be quickly analyzed and cross-checked. In Table 1, the column titled "0 Years Since Launch" in effect represents the launch date of the drug in question, and as such should have no Prescribers at that particular

* As a practical matter, the choice of treatment option inclusion often replicates the mutually exclusive and collectively exhaustive set defined in the Patient Flow portion of an integrated dynamic model (see chap. 4).

† The choice of stopping this longitudinal look at 5 years postlaunch is purely for illustration purposes. Most drugs can and do have a much longer lifecycle.

‡ For example, prescribing behavior in the anti-depressant markets is no longer the exclusive domain of psychiatrists.

Table 2 Computed Values for Coefficients Related to Internal and External Influence in Example Doctor Adoption Framework

Years of launch	1	2	3	4	5
Internal influence	0.1	0.1	0.1	0.1	0.1
External influence					
Treatment 1	0.1	0.25	0.35	0.3	0.2
Treatment 2	0.05	0.06	0.1	0.1	0.12
Treatment 3	0.05	0.1	0.25	0.25	0.2

time point. There could, however, be prelaunch Awareness established at the time of launch.

The metrics shown in Table 1 allows the stock/flow framework and associated factors that govern adoption of these drugs to be calibrated to historical data. The aggregate metrics can be combined with the described structure, and with the help of optimization algorithms in Dynamic Modeling software, the relative importance of *internal and external influences* in the diffusion process can be determined. These influencers are usually described as coefficients, and the results of this calibration process result in specific computed values for these variables, as shown in Table 2.

Note that in our example case, the derived value for *internal influence* is constant over time and the same for all treatment options in the mutually exclusive and collectively exhaustive set. This is not true by definition, but instead provides a useful starting point for the calibration. An extension of this basic approach might involve having *internal influence* be specific for each product. This word-of-mouth factor is often expressed as a function of overall product attractiveness, such that very good products might have a high word-of-mouth coefficient in a positive direction while very bad ones just the opposite. An alternate approach is to make the *internal influence* variable change over time, such that later drug releases might have a hard time generating word of mouth given crowded marketplace conditions. This more detailed option may be useful, but we have found that in most markets, assuming a constant value for *internal influence* is a good first approximation.

By including these factors into the dynamic stock/flow framework for Doctor Adoption, the simulation outputs closely match the aggregate market metrics for Awareness and Prescribing as shown in Table 1. All models must define an acceptable error level during the calibration process, and dynamic models are no exception. While analytic computations such as sum of squared errors (SSE) or other goodness of fit metrics are beyond the scope of this work, they can certainly be applied to ensure statistical validity to simulation results.

Note also in Table 2 that the *external influence* coefficients vary over time by product. Again, this is not necessarily the case, but is often true in the markets we have examined. One value of taking this approach is that the factors that determine the aggregate measures of *external influence* (detailing, sampling, etc.) can now be explored.

EXPANDING THE DYNAMIC MODEL TO INCLUDE CAUSAL FACTORS OF *EXTERNAL INFLUENCE*

Various components of promotional spending can be explicitly included and analyzed for a particular marketplace by expanding the basic Doctor Adoption component of the Standard Template. This type of extension makes apparent the goals of any type of physician marketing and leverages the knowledge of the cross-functional team in understanding the dynamics involved in the adoption process. The Doctor Adoption framework is ultimately combined with the other sectors of the Standard Template to arrive at an integrated model of the market from which metrics such as ROI can be calculated. It starts, though, with an explicit rendering of elements of promotional spending into the stock/flow adoption framework.

As we noted earlier in this chapter, *external influence* is an amalgamation of those factors under the control of a pharmaceutical company actively marketing a particular therapy in an indication. The types of marketing activities range from one-on-one with doctors (detailing) to mass market messaging (televised advertising.) In our Standard Template, however, we generally limit our choices to broad categories for which data is readily available and can be easily "rolled up" to the *external influence* factor calculated by aggregate analysis. To extend our continuing example, assume the following marketing categories have been present in the launch dynamics for the treatment options in our sample market:*

- Detailing = Sales representatives spending time with individual doctors
- Samples = Money spent on free trial samples to physicians
- Medical Education = Large symposia attracting prescribing physicians

These factors can be incorporated into the Standard Template by extending the basic structure, as shown in Figure 8.

An example of product-specific spending numbers for each postlaunch year is shown in Table 3.

Recall that we are looking at the data shown in Table 3 *only* for an attempt to discover the relative benefits of various types of marketing spending in this arena. For example, note that total marketing spend for Treatment 1 and Treatment 3 is identical in each year of launch, but their respective calculated *external influence* in the Doctor Adoption process varies greatly. Conversely, Treatment 2's overall marketing spending (in this example) is the same for the first five years of launch, but its computed *external influence* factor shows wide variation. Each of these examples indicates that changes in the marketing mix are driving these differences in observed behavior. It is only these variations that drive us to the

* Other marketing variables can be easily included in this framework, but will not be shown in this example for the sake of simplicity.

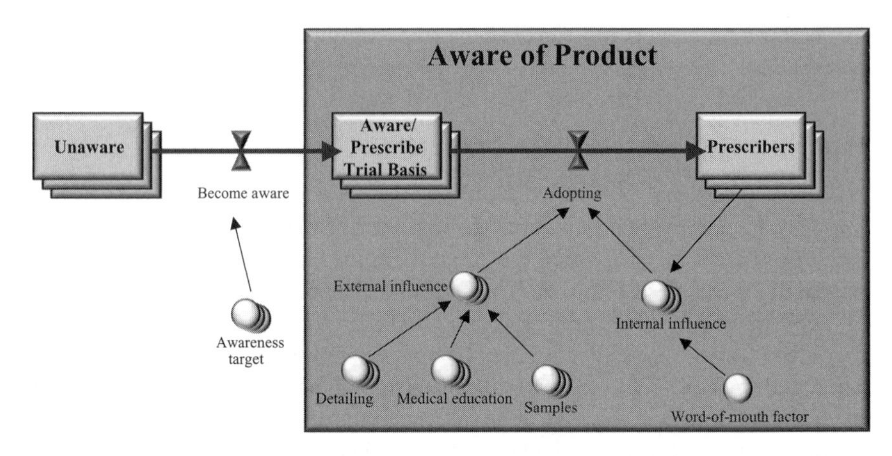

Figure 8 Expanded Doctor Adoption framework including factors of *external influence*.

level of the individual components of marketing spending. Had the data showed a more consistent relationship between overall marketing spending and its expected impact on diffusion, this degree of disaggregation may not have been necessary.

However, there exists a body of research in this arena that often proves useful in deriving the effects of various marketing levers on Doctor Adoption (3,4). These studies provide a statistical methodology help place bounds on certain parameters governing the diffusion of pharmaceutical products through the prescribing physician universe. The mathematics and analysis contained in these algorithms is beyond the scope of this work, but are excellent sources for the interested reader. By following these established approaches from the world of operations research and integrating them into the Dynamic Modeling framework, the relative importance

Table 3 Promotional Spending Numbers for Treatments Options in Example Marketplace

Years of launch	1	2	3	4	5
Total Dr. Marketing spend: Treatment 1 ($)	20	40	50	40	30
Detailing ($)	10	20	30	30	20
Samples ($)	4	10	10	5	5
Medical education ($)	6	10	10	5	5
Total Dr. Marketing spend: Treatment 2 ($)	15	15	15	15	15
Detailing ($)	3	4	7	10	10
Samples ($)	4	5	5	3	3
Medical education ($)	8	6	3	2	2
Total Dr. Marketing spend: Treatment 3 ($)	20	40	50	40	30
Detailing ($)	4	4	15	20	20
Samples ($)	6	6	15	10	5
Medical education ($)	10	30	20	10	5

Table 4 Derived Coefficients for Factors of External Influence in Doctor Adoption Structure

Variable	Calculated coefficients
detailing	0.01
samples	0.004
medical education	0.001

of each component in the marketing mix can be calculated. Such importance metrics are often expressed as coefficients (similar to the coefficients of *internal* and *external influence*) to show relative strengths of the input parameters. In our continuing example, the coefficients can be calibrated using the dynamic model and are shown in Table 4.

Without getting too caught up in the mathematics, the relative size of each coefficient in Table 4 shows the importance of that component of physician marketing spend on the overall diffusion process in our example marketplace. In this sample case, spending on *detailing* appears to have the most impact on Doctor Adoption—its calculated coefficient is the highest. In fact, the derived impact of *detailing* is 2.5 times that of *samples* ($0.01/0.004 = 2.5$) and 10 times that of *medical education* ($0.01/0.001 = 10$) in this example marketplace.

We can now extend the Doctor Adoption model to include historical marketing spending from Table 4 to calculate the resulting impact on diffusion of treatment options, calibrated to historical data on Prescriber status. The results of this expansion and the corresponding dynamic model are shown in Table 5.

Not only do the dynamic structure and the associated data/calibration shown in Table 5 explain the historical behavior of the market very well; it provides a useful tool for analyzing the expected results of a newly released drug into the marketplace. For instance, we could extend this dynamic model to include a slot

Table 5 Comparison of Dynamic Model-Generated Results vs. Historical Data in Example Indication Marketplace

Years since launch	0	1	2	3	4	5
Model results: Prescribers						
Treatment 1	0	70	90	95	98	99
Treatment 2	0	0	3	15	27	29
Treatment 3	0	9	72	97	100	100
Historical data: Prescribers						
Treatment 1	0	70	90	95	98	100
Treatment 2	0	1	3	15	25	30
Treatment 3	0	10	70	95	98	100

Table 6 Example Targets for Awareness and Marketing Spending for Treatment 4 in Sample Indication Marketplace

Years of launch	0	1	2	3	4	5
Expected awareness: Treatment 4 (%)	20	30	40	45	50	50
Total Dr. Marketing spend: Treatment 4 ($)		**40**	**50**	**30**	**25**	**20**
Detailing		20	30	20	15	10
Samples		10	10	5	5	5
Medical education		10	10	5	5	5

for a soon-to-be-offered drug (call it Treatment 4, from our continuing example) to estimate how rapidly it would be adopted, given the derived values governing historical adoption of competitive products. Goals or forecasts of expected awareness metrics and marketing spending levels might be established for each of the first five years of Treatment 4's launch, as shown in Table 6.

Including these spending and awareness scenarios for Treatment 4 as shown in Table 6 into the dynamic model gives the results shown in Figure 9.

The projection of expected prescriber status shown in Figure 9 provides a useful base case scenario from which various strategic options can be rigorously tested, such as the following:

- Would higher awareness early in the launch of Treatment 4 result in faster adoption on the part of prescribing physicians?
- How would changing in the expected marketing mix affect the speed and magnitude of the diffusion path?

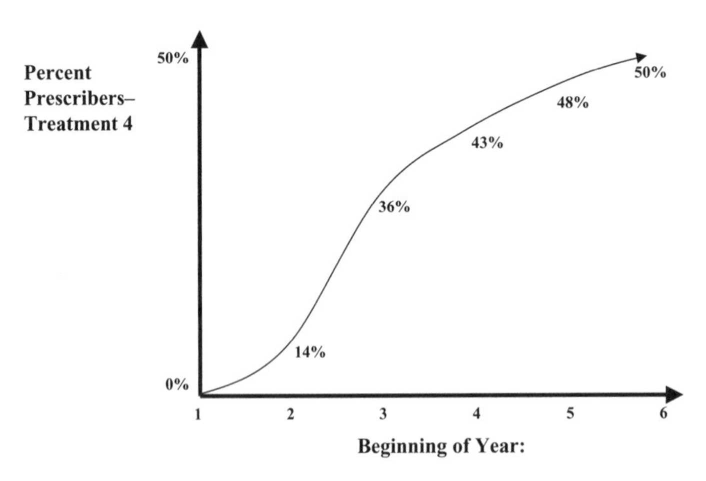

Figure 9 Example dynamic model projection of Doctor Adoption of Treatment 4.

- How important is word of mouth (the auxiliary variable *internal influence*) in the adoption process, and are there any strategic levers to leverage that dynamic for Treatment 4?
- How important is the overall prescriber status when this sector of the dynamic model is integrated with other components of the Standard Template?

These questions, framed and ultimately answered using the Dynamic Modeling framework to approach the issue of Doctor Adoption, can be of extreme importance in establishing an effective Brand Plan for the launch of a pipeline pharmaceutical product. The strategies included in brand positioning of the compound can subsequently be based on the knowledge of the Doctor Adoption process as discovered through the modeling exercise, providing a means to more effectively allocate marketing resources to area of high leverage.

Strategic Implications of Dynamic Modeling for Doctor Adoption

It is no secret that the diffusion of newly released pharmaceutical compounds throughout the population of prescribing physicians has a huge impact on the ultimate performance of an NCE, but traditional approaches often fail to explicitly capture this important dynamic. Such adoption behavior is usually included as an implicit part of the expected sales trajectory for a compound or at best as an input scenario defining the "time to peak sales" and the shape of that ascent. There is rarely a clear understanding of how various marketing strategies toward physicians might affect the expected sales trajectory, or to what extent such dynamics are influenced by internal factors such as word-of-mouth or external factors such as sampling, detailing, clinical trials, journal advertising, etc.

For example, one of our clients had a significant hole in its portfolio due to a number of their key compounds coming off patent, and the resulting loss of revenue was expected to have a huge effect on the firm's financial stability. An NCE for a major therapeutic area was close to finishing Phase III trials, and its early clinical profile showed it to be a major breakthrough in treatment of a particular disease. The interest in this new compound went beyond the Brand Planning team to an executive committee created to evaluate the commercial potential of the NCE. This new compound needed to have an uptake that was significantly faster than that of previously released compounds in the indication in order to plug the revenue gap looming on the horizon.

First, a dynamic model of the Doctor Adoption process was created and populated with historical data from existing drugs in the indication to get a sense for the drivers of diffusion in the disease marketplace. A calibration process determined the relative effectiveness of physician marketing initiatives based on these historical analogs. Next, this diffusion structure was combined with Patient Flow and Treatment Attractiveness sectors to arrive at an integrated model of marketplace dynamics. Finally, the resulting simulation tool was used to test various levels of marketing, spending, and physician programs needed to achieve the desired penetration and resulting revenue streams necessary from a strategic perspective. The Dynamic Modeling process and resulting model allowed the executive committee to ask and answer questions such as the following ones:

- How many physicians are necessary at the launch and in the short term to meet the financial objectives for the compound?
- How can resources be allocated across various types of physician marketing initiatives in order to achieve those metrics in the most cost-effective manner?
- How should strategies be sequenced in order to maximize performance in the marketplace?
- How do various levels of marketing spending correspond to expected sales trajectory for the compound?
- How big are the various segments of prescribing physicians, how different is their adoption behavior, and what segment-specific strategies should be employed?

The comprehensive look at the Doctor Adoption process allowed the client to systematically evaluate its strategic options, and ultimately formulate a series of strategies that was robust enough to achieve its financial targets.

An understanding of the structure and behavior of the diffusion of pharmaceutical products through the physician universe is essential for formulating effective marketing strategies for NCEs as they enter the marketplace. The Dynamic Modeling approach also provides operational rigor into commercial evaluations of a compound's financial potential by explicitly capturing and analyzing key Doctor Adoption dynamics. Models of such diffusion patterns range from the simple to the very complex, but the basic organizing framework of the Dynamic Modeling methodology can easily be adapted depending on the needs of the Brand Planning team, the complexity of the therapeutic area, and the level of specificity required.

SEGMENTATION

Another extension of the basic Doctor Adoption framework is the inclusion of segments of prescribing physicians. Generally, such segmentation takes place on one or two dimensions:

1. *Specialty* = General Practitioner/Primary Care Physician vs. Indication Specialist.
2. *Adoption rate* = Doctors categorized according to their proclivity to adopt newly released pharmaceuticals into their prescribing patterns.

Disaggregation across a Specialty dimension is usually straight forward, both in terms of quantification of target segments ("How many PCPs treat this

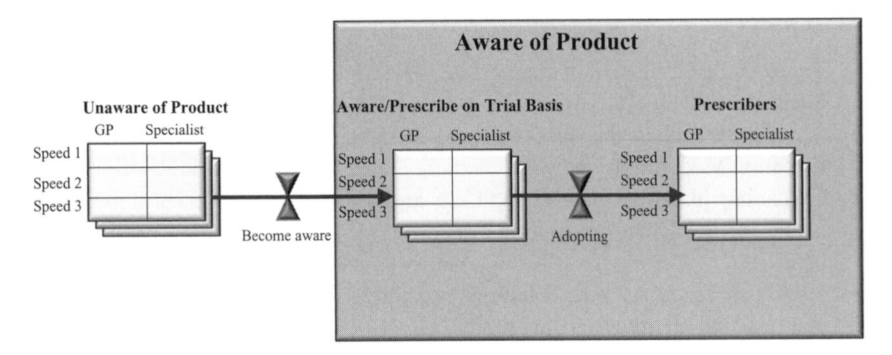

Figure 10 Stock/flow framework for Doctor Adoption disaggregated by specialty and adoption rate dimensions.

disease, and what percent of the prescriptions come from them?") and in the factors that govern them ("How many detailing dollars were spent by Treatment 1 on Specialists in 2008?"). Generally, such data on this dimension of segmentation are readily available, either internally or through various second-hand vendors.

Details on Adoption Rate tend to be more difficult to obtain. Conceptually, disaggregation on this dimension attempts to segment the target physician population according to propensity to adopt new products. This approach to segmentation has been investigated for many years and has a rich tradition in the published literature.* For this disaggregation approach to be of value in our dynamic stock/flow framework, the elements of the dimension must be determined, and the subsequent differential values dictating each segment's adoption must be analyzed. To do so, aggregate metrics of Awareness and Prescribing must be available for each target segment ("10% of Speed 1 doctors were prescribing Treatment 2 three years post-launch compared to only 5% of Speed 2 doctors.") Typically, such market-level metrics based on adoption proclivity are difficult to obtain, making this disaggregation an interesting intellectual (but difficult to quantify) modeling exercise. However, conceptually, such a stock/flow segmentation looks something like what is shown in Figure 10.

As we saw in chapter 5, the introduction of disaggregations into a dynamic model is easy to implement but often difficult to quantify. The caveats presented in that chapter regarding Patient Flow complexity apply to the Doctor Adoption realm as well. However, some of our clients have been able to support these types of disaggregations with associated data that provided additional market insight into the dynamics of diffusion of newly released treatments.

* One popular example is Geoffrey Moore's *Crossing the Chasm: Marketing and Selling High-Tech Products to Mainstream Customers* (6), in which he details the attitudes and associated marketing strategies to address Innovators, Early Adopters, Early Majority, Late Majority, and Laggards in technology markets.

SUMMARY

- There is a rich history of mathematic approaches to analyzing new product diffusion.
- Algorithms such as the Bass Diffusion model can easily be translated into the Dynamic Modeling framework.
- Collecting historical data on ATU in an indication marketplace allows the results of the Doctor Adoption framework to be calibrated to past market behavior.
- Marketing levers such as detailing, sampling, and medical education can be incorporated into the dynamic model.
- The resulting simulation tool is a useful scenario analysis tool to develop effecting marketing strategies.
- The Doctor Adoption structure can be disaggregated to include identified segments of prescribing physicians.

REFERENCES

1. Gagnon MA, Lexchin J. The cost of pushing pills: A new estimate of pharmaceutical promotion expenditures in the United States. PLoS Med 2008; 5(1).
2. Liebman M. Finally, Predictive Returns on Promotional Investments. Medical Marketing and Media, 1998.
3. Berndt E, Pindyck R. Network Effects and Diffusion in Pharmaceutical Markets: Antiulcer Drugs, NBER working paper No. 7024, March 1999.
4. Zoltners A, Hahn M, Park S, et al. Analysis of New Product Diffusion Using a Four-Segment Trial Repeat Model", Marketing Science, Summer 1994.
5. Sterman J. Business Dynamics: Systems Thinking and Modeling for a Complex World. Boston, MA: McGraw Hill, 2002.
6. Moore G. Crossing the Chasm: Marketing and Selling High-Tech Products to Mainstream Customers. HarperBusiness, 1999.

9

Agent-Based Modeling Approach

INTRODUCTION

This chapter will cover the following topics:

- Situational need for more sophisticated modeling approach
- Explanation of agent-based modeling (ABM) methodology
- Examples of dynamic models incorporating agent-based approach
  - Therapy duration
  - Changing segmentation thresholds
  - Overlap of multiple attributes
  - Line of therapy (LOT) dynamics
  - Limited number of patients in the dynamic model
  - Detailing and word of mouth (WOM) in Doctor Adoption framework
- Strategy tests using agent-based models
- Current software tools

With increased breadth and depth of patient-level databases and information regarding physician prescribing behavior, the issue for most Brand Planning teams in the pharmaceutical industry is not the scarcity of data but rather the lack of effective tools to fully utilize and leverage it. As detailed in previous chapters, traditional Dynamic Modeling goes a long way toward solving this problem by providing a simple yet powerful framework, which can be extended and augmented to meet the vast majority of analytical situations. In certain cases, however, we have found the need to expand the boundaries of the Dynamic Modeling toolkit (which is based on traditional System Dynamics principles) and utilize the more sophisticated approaches from the world of computational science. One such methodology that we increasingly employ within the broad Dynamic Modeling construct is a technique called ABM.*

* Although we feel the term "agent" is not entirely illustrative for nontechnical folks, the phrase is now standard and hence part of our modeling vocabulary.

ABM departs from the more aggregate, top-down approach as illustrated thus far in this work and instead addresses the behavior of individual items (people, doctors, development compounds,* etc.) within a system. In other words, ABM creates simulation entities which can be given specific characteristics, behavioral parameters, and/or rules of interaction.† The ABM approach comes very close to the evolving concept of "virtual reality" in which a computer environment is used to replicate the complex series of interactions on the part of individuals, which produce a wide variety of collective behavior. Use of ABM in the pharmaceutical realm is not focused on these theoretical notions, but rather as a means to solve problems in situations where traditional methods struggle to produce meaningful results. In particular, ABM is often warranted if one or more of the following conditions are met:

- The number of things to be tracked in the simulation [commonly patients, physicians, New Chemical Entities (NCEs)] is relatively small
- Significant segmentation has reduced previously large groupings to very small number of items in each category
- Items in the simulation are to be tracked by a large number of characteristics or dimensions
- It is difficult to determine useful disaggregation categories based on established, unchanging thresholds
- Overlap of disaggregation categories (such as comorbidities) is challenging from a data collection standpoint
- Traditional disaggregation approaches result in very slow simulation models
- There are significant interaction effects between simulation items that cannot be addressed in a basic System Dynamics framework

EXAMPLES OF ABM APPLICATIONS

In our client work, we have found that a few simple examples can illustrate the concept behind ABM more effectively than detailed, technical explanations. The example cases which follow illustrate how the ABM methodology is often employed.

Therapy Duration

Chapter 6 introduced the concept of probabilistic models which in a traditional framework can be used to incorporate model inputs pulled from a random

* A detailed example of the use of ABM to analyze pipeline and portfolio dynamics is the focus of chapter 16.

† Readers may be familiar with early computer simulations involving ABM such as "Game of Life" or "Boids" in which simulation entities or agents were given very simple behavioral rules which collectively result in interesting aggregate patterns of behavior.

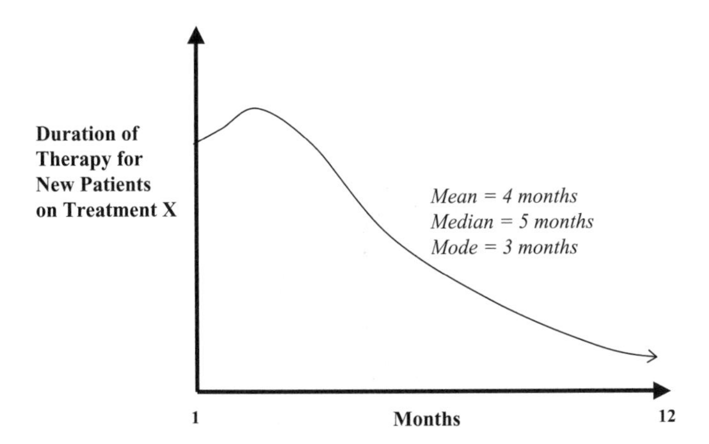

Figure 1 Example new patient therapy duration—Treatment X.

distribution. Recall that with a typical disaggregation structure, collections of patients can be subcategorized into smaller groupings based on any of a number of characteristics. Where the ABM approach differs from this traditional methodology, however, is in the ability to treat each individual patient as a separate and distinct entity, with attributes assigned and pertaining to that specific individual. This flexibility turns out to be incredibly powerful in addressing issues such as ranges of therapy durations for a given patient population.

The example in chapter 6 regarding therapy durations presented a sample case in which the expected length of time that new patients would stay on Treatment X was not a fixed number, but rather was distributed across a range of possible values (see Fig. 5). With the increased sample sizes available from many longitudinal integrated claims (LIC) databases, it is often possible to collect a large amount of data regarding therapy durations for various groups of patients. With this information in hand, statistical programs such as SPSS® or SAS® can derive the implied distribution represented by the historical data, as shown in Figure 1.

The derived distribution of therapy durations allows the ABM to tag each individual patient (represented by a simulation entity) with its own specific and precise input parameter regarding precisely how long that individual patient will remain on Treatment X. In other words, the ABM framework allows us to depart from the concept of average durations of therapy that must be applied to a particular group of patients (in either deterministic or probabilistic fashion) and instead assign a specific therapy duration to a stand-alone simulation agent That simulated patient will then remain on therapy for exactly the period of time indicated by its assigned therapy duration—which is a parameter pulled from the derived distribution of the length of time patients stay on Treatment X. In effect, the ABM is recreating the individual details of the dynamics surrounding the number of patients on Treatment X by treating patients as distinct individuals, as shown in Figure 2.

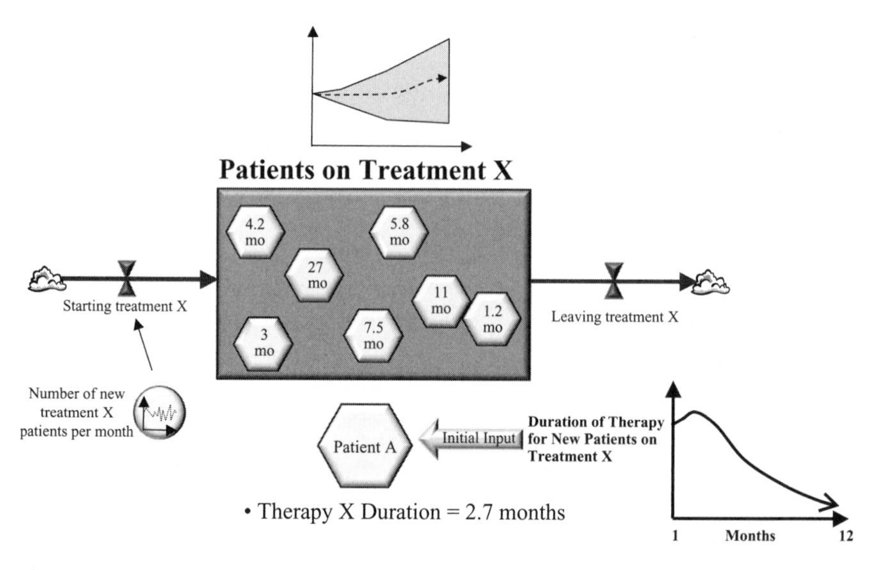

Figure 2 Example ABM structure and output—therapy duration.

ABMs such as the one depicted in Figure 2 will display a range of possible outcomes, because in each simulation run every individual patient is assigned a particular therapy duration from a derived distribution. Multiple simulation runs must then be aggregated and compared in order to assess possible variation in outcomes, given variability in the input assumptions. More importantly, from a Brand Planning perspective this type of ABM comes very close to replicating the real-world through the use of computer simulation, providing a great tool not only for analysis but also scenario planning and strategy development.

Changing Segmentation Thresholds

Suppose that a marketing team is interested in analyzing treated diabetic patients by categorizing them according to body mass index (BMI). One way to accomplish this task would be to choose appropriate "cutoffs" for BMI, establish appropriate patient groupings (see chaps. 5 and 6), and simply distribute the number of diabetic patients according to published research and/or analysis of patient databases. For example, the team might decide appropriate buckets of patients include those with BMI <30, BMI from 30 to 40, and BMI >40. This approach is illustrative of a traditional disaggregation or segmentation process, as we have described previously. This set bucketing process has certain drawbacks, however. First, although these defined categories might be meaningful in the current timeframe, in the future it is possible that other thresholds (perhaps determined by changing guidelines regarding BMI might be more appropriate. Second, once these categorizations are set in a traditional segmentation framework, the ability to analyze any different "cuts" is lost. In other words, once patients are bucketed into the BMI <30 category, it is

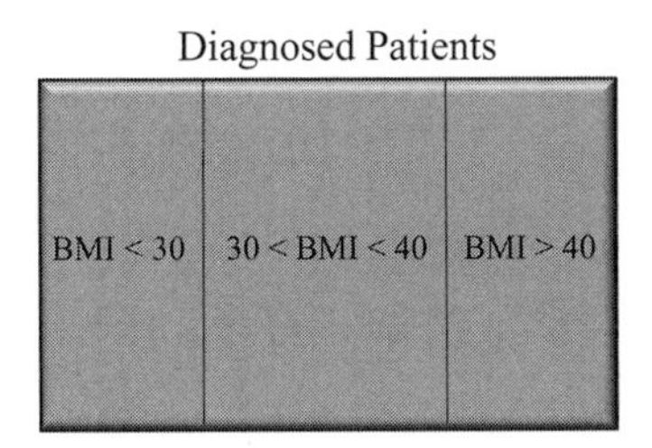

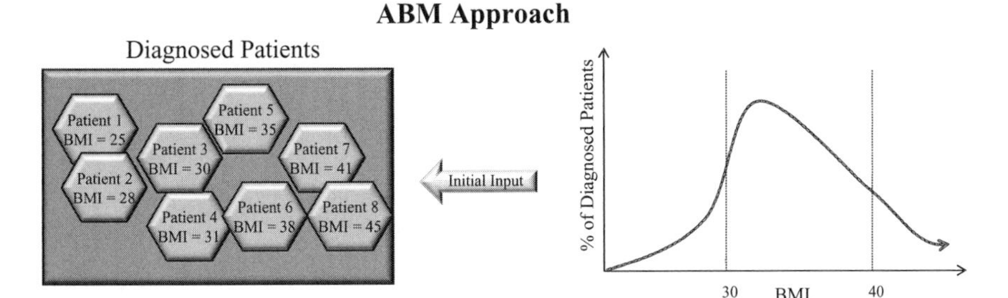

Figure 3 Traditional segmentation vs. ABM approach—BMI example.

no longer possible to see the number of patients with a BMI <25, for example, or calculate how many patients have a BMI between 27 and 29.

In an ABM framework for BMI, set segmentation buckets do not need to be established a priori. Instead, the model would be created and initially populated with a number of simulation agents with a characteristic or attribute called "BMI." Each individual simulation agent would be given an initial BMI whose value is drawn from a distribution of BMI for treated diabetic patients, as shown in Figure 3.*

Next, each simulation agent (representing an individual patient) would be given behavior rules concerning changes in BMI over time. These rules can be deterministic ("the BMI for each patient will increase by 1 for each year of simulated time") or probabilistic† ("10% of patients will have their BMI increase by 2 each year, 70% will have their BMI rise by 1 per year, and 20% of patients will experience a decrease of BMI of 4 over a period of 5 years.") In fact, the ABM approach gives the model designer an incredible amount of freedom regarding the rules governing the simulated behavior of the agents/individuals in the system.

Once populated, the simulation can be run and the results analyzed. As was detailed in chapter 6, if the model has probabilistic elements, multiple simulation

* For simplicity, we are showing only eight individuals, but current ABM software has the ability to handle thousands and even millions of simulation agents.

† See chapter 6.

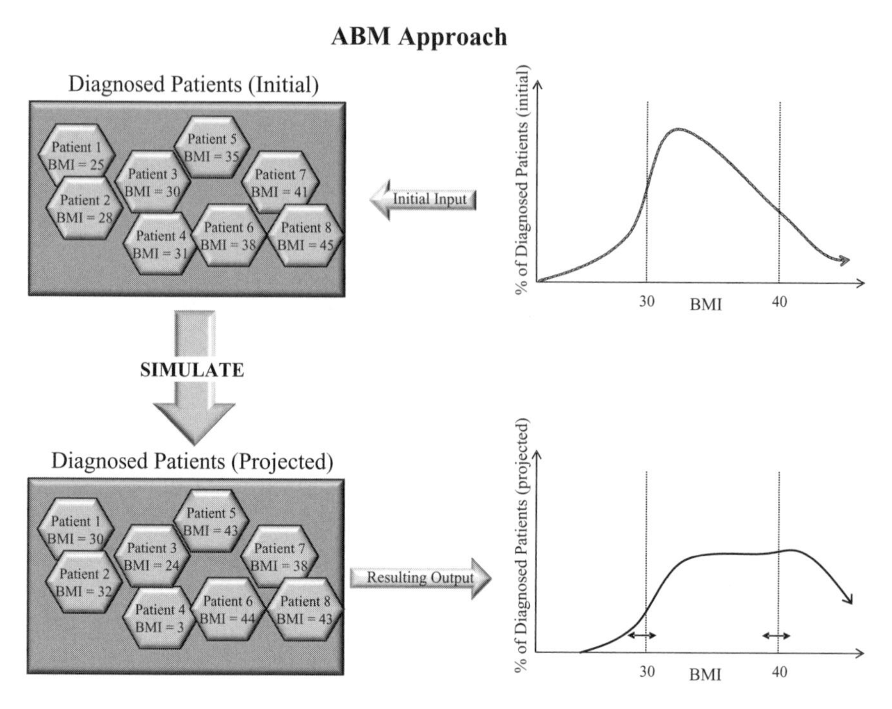

Figure 4 ABM initialization, simulation, and projected outcomes—BMI example.

runs will have to be performed in order to correctly interpret results. However, the advantage of the ABM approach is that since each individual simulation agent is holding its own specific BMI throughout the simulation the aggregate results can be collected and "bucketed" in whatever categories seem relevant.

Suppose that the team wanted to look at what the BMI distribution might look like 10 years in the future, given the current (existing) BMI distribution and incorporating the behavioral decision rules regarding BMI evolution for individual patients. For the sake of simplicity, we will assume that no other dynamics (treatment discontinuation, reinitiation, and/or mortality) are at play within the model—so we are simply looking at how the distribution of BMI is changing for a set cohort of patients. Depending on the behavioral rules for patients in the simulation regarding their BMI, the projected distributions of patients might look vastly different from the initial ones. In addition, these patients can now be "sliced and diced" according to any desired BMI category, as depicted in Figure 4.

The degree of flexibility offered by an ABM approach, as demonstrated in this simple example, is often beneficial to a Brand Planning team whose mission is to look not only at what is happening in today's marketplace, but also to develop effective strategies for how the market will look in the future.

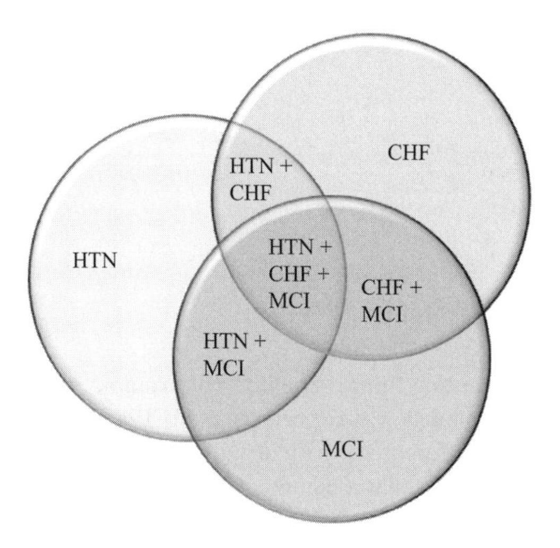

Figure 5 Overlap of comorbidities—diabetes example.

Overlap of Multiple Attributes

Issues of comorbidities often plague traditional patient segmentation schemes, often because published research or even careful analysis of International Classifications of Disease (ICD-9) diagnosis codes in databases is insufficient to correctly categorize overlap of existing conditions. To continue the example from above, suppose in addition to segmenting patients according to their BMI, a marketing team is interested in bucketing diabetic patients according to a set number of comorbid conditions such as hypertension (HTN), congestive heart failure (CHF), an experience with a myocardial infarction (MCI), etc. Typically, overlap of comorbid conditions can be analyzed and ultimately displayed by a Venn diagram, as shown in Figure 5.

Layering in another dimension (BMI) into this patient disaggregation can present significant difficulties from a data collection standpoint. Will it be possible to further segment the patients having HTN + CHF + MCI according to their BMI category? And to the point demonstrated in example 2, what if the guidelines for HTN ultimately change and/or the team wants to further break down patients into classes (I–IV) of CHF? From a traditional segmentation standpoint, this process quickly becomes unwieldy.

An ABM approach would tackle this problem differently. Instead of defining set patient segments and their explicit overlap, agents (representing patients) would be given attributes related to the categorizations desired by the marketing team, as shown in Figure 6.

As described previously, each simulation agent/patient would take on initial parameters based on observed patient distribution for each attribute/characteristic. As time progresses through the simulation, each patient's attributes would be

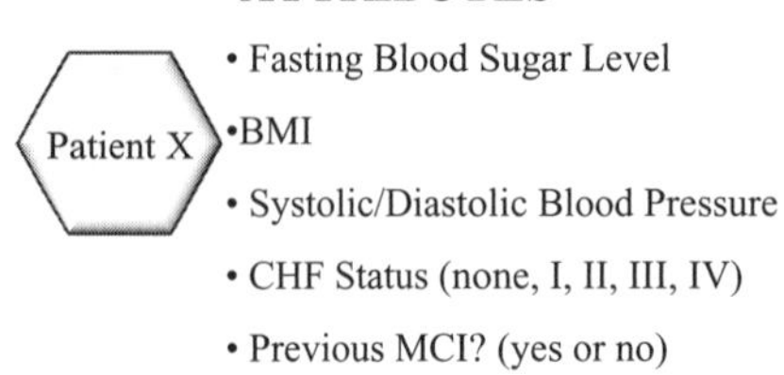

Figure 6 Simulation agent and associated attributes—diabetes example.

updated based on programmed rules regarding future behavior. For example, what is the probability that a CHF Class III patient will experience a MCI in a given year? How are blood pressure metrics different for patient with HTN vs. those with just a BMI >40? What is the likelihood that a diabetic Class I CHF patient will deteriorate to a Class II categorization each year? When these behavioral rules and integrated and then simulated, the characteristics of the patient population (as represented by simulation agents) will evolve over time. In this manner, the ABM approach handles multiple patient attributes in a much more elegant and efficient way, allowing the Brand Planning team flexibility and added analytical rigor. In addition, the simulation outputs from an ABM can be organized according to these specific patient attributes and then displayed using pivot tables in a spreadsheet format, providing an incredible amount of reporting flexibility. Many of our clients use such pivot tables to effectively "slice and dice" the results from an ABM simulation along a number of different categorization dimensions.

LOT Dynamics

Recall from chapter 5 that the stock of Currently Treated patients in the Standard Template can be expanded to include LOT dynamics. For example, instead of simply calculating the probability of patients switching treatments, the Standard Template can be organized such that it keeps track of patients based on their sequential treatment choices (see subsection "Case Study: Lines of Therapy Dynamics" in chap. 5). For a small number of possible therapy options and a minimal number of LOTs, the approach outlined in chapter 5 works reasonably well. But increasingly, Brand Planning teams want to track a large number of treatments and follow patients as they progress through multiple (up to 10, in some cases) LOTs. Doing so begins to present a problem from a standard disaggregation standpoint, as the resulting dynamic model often becomes heavily segmented and inefficient. To understand the reason, suppose a marketing team is interested in tracking 10 different treatment options and 8 LOTs. In effect, this creates 80 different patient segments (10 therapies × 8 lines). But each individual patient can only take one path through this treatment progression—there are of course 80 possibilities, but in a traditional segmentation structure the dynamic model must track the past treatment choice options as well the current position of a patient. For example,

Table 1 Example Treatment History for Hypothetical Patient on 4th LOT

LOT 1	LOT 2	LOT 3	LOT 4
A			
	B		
			C
		D	

consider a patient who is on Treatment C and is in the fourth LOT, with the treatment history as outlined in Table 1.

As is evident in Table 1, there are eight distinct data parameters related to the patient in this example. These data detail which treatment the patient was on in each of four LOTs. However, in a traditional disaggregation framework, the segmentation would also have to account for all the treatments the patient *was not on* in each of his/her previous LOTs, as well as those he/she *might be on* in future LOTs. That means that for each slice of simulated time, all 80 cells of the treatment/LOT grid need to be calculated, even though the vast majority of those cells (90%, to be exact) are not really "in play."

Computer science experts would call such a situation "computationally inefficient." We would call the resulting simulations "really slow." Because the traditional disaggregation framework is forcing the simulation to make many more calculations than is actually necessary, resulting simulation runs often seem interminable. Even with advanced computing power, slow simulations rarely make effective tools for either analytic types or for Brand Planning teams seeking a more effective decision-making process through the use of simulation.

In contrast, an ABM approach to the LOT problem simply assigns an attribute to a simulation agent (again, representing a patient) concerning the therapy choices. These choices are incremented and accumulated as the patient moves through an LOT progression. When Treatment A was initiated in LOT 1, that information would be a characteristic "attached" to that agent. Subsequent treatment transitions would be coded in a similar fashion, as shown in Figure 7.

In this fashion, the ABM framework provides an efficient way to collect information on a patient's current therapy use as well as treatment history, which opens the door for some quite sophisticated LOT dynamics. For example, previous chapters have discussed the concept of Treatment Change Matrices, in which the

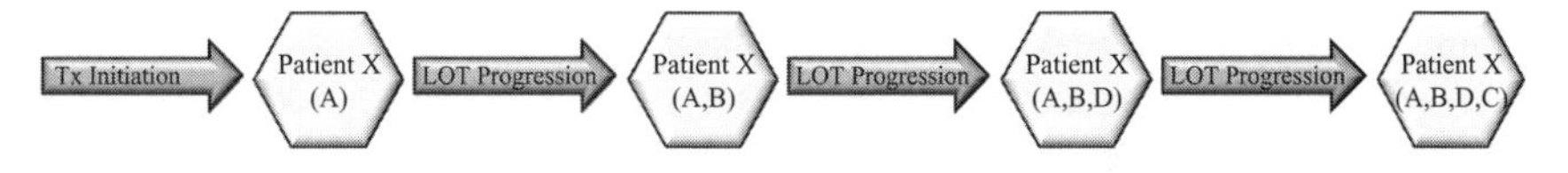

Figure 7 Example therapy progression and accumulated treatment history using ABM approach.

probability of moving to the next therapy regimen is dependent on the current treatment option. These transition probabilities are known in mathematical circles as a first-order Markov process in which the "order" of the Markov model refers to how far back in the decision-making process (in this case, a patient's treatment history) we choose to go. In other words, a Markov process says that a patient's next therapy choice is a conditional probability. In a first-order Markov model, the only relevant condition is the patient's current therapy choice. In a second-order Markov model, for example, the transition probabilities would be a function not only what treatment the patient is currently on, but also what his/her most recent therapy choice was. The ABM methodology provides an elegant way for *all* of a patient's treatment history to be incremented and collected, so that subsequent choices regarding therapy option may be appropriately determined based on careful analysis of historical data. This advanced method for handling LOT progressions is of particular importance in two different cases:

1. In markets (such as HIV) in which patients rarely, if ever, return to a particular therapy after transitioning from it.
2. For indications where the discontinuation of one drug in a particular class at any point in a patient's treatment history precludes returning to another new drug in the same class down the road.

Figure 8 shows an example of how an ABM would handle the transition probabilities for two hypothetical patients based on cumulative treatment history.

As marketplaces for pharmaceutical products become increasingly crowded, the ability to analyze the specifics of patient behavior using LOT dynamics has become even more vital for strategic success of new compounds. By leveraging historical data and employing an ABM approach, a Brand Planning team can utilize this sophisticated approach to more effectively formulate successful marketing strategies.

Limited Number of Patients in the Dynamic Model

Many pharmaceutical products are developed to treat diseases afflicting millions of patients, and representative databases for these populations tend to provide statistically significant samples from which data regarding patient behavior can be extracted. However, certain indications have very small patient populations, and simulating the behavior of those patients using a traditional Dynamic Modeling framework can be challenging. Particularly when the aggregate patient population is segmented (into age or gender categories, for example), common modeling frameworks begin to break down and lose their usefulness.

ABM is well suited to address this type of situation, however, and can establish a distinct agent for each individual patient whose subsequent behavior is treated probabilistically. For example, suppose a pharmaceutical company is developing a compound to be used by patients following a pancreas transplant.

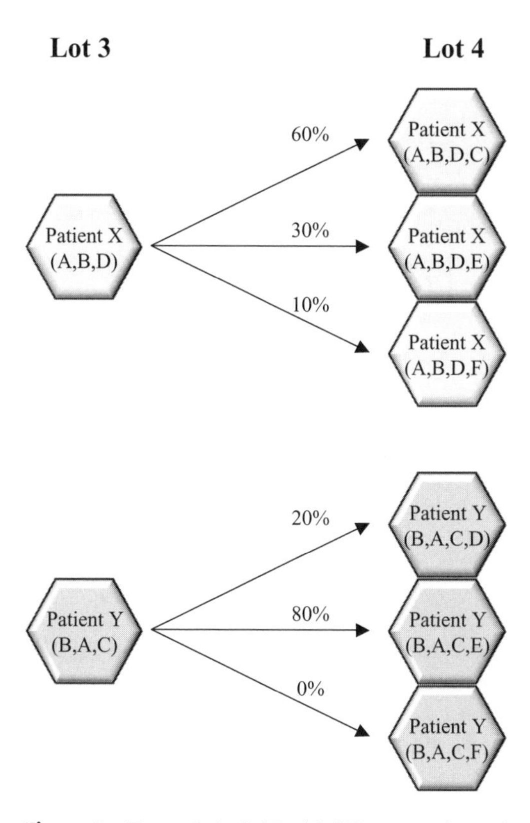

Figure 8 Example individual LOT progressions given cumulative treatment histories.

It is expected that patients would use the therapy for up to 3 years following the operation to prevent organ rejection, and is likely to have an increased mortality benefit of approximately 10%. The Brand Planning team is interested in forecasting the total number of patients who might be using the drug, even though there is a high degree of uncertainty related to how many patients will be placed on the drug, how long patients will use the drug, the true impact on survivability, etc.

The Organ Procurement and Transplant Network (OPTN) reports that 469 pancreas transplants took place in the U.S. in 2007, and OPTN also publishes survival rates for the 1-, 3-, and 5-year marks (along with confidence intervals.) These data could be incorporated into a very simple Dynamic Modeling framework showing the relevant stocks/flows of patients, but populated with specific simulation agents rather than a basic numerical representation. In effect, ABM simulation would create simulation agents (representing individual patients) as they start on this new drug. Individual patients would also be assigned attributes related to how long they are expected to take the treatment as well as how long they are expected to live after the pancreas transplant. Couple these parameters with variability around the number of projected pancreas transplants as well as

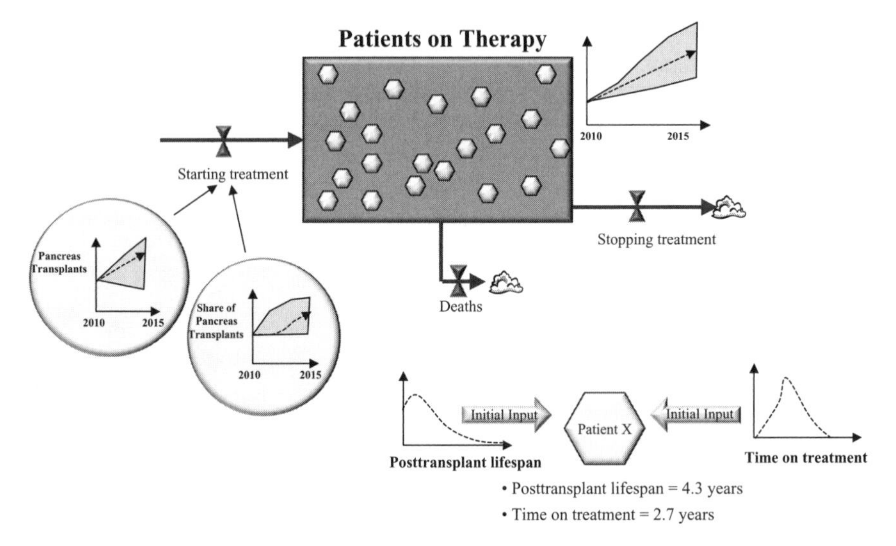

Figure 9 Sample inputs to and outputs from ABM for pancreas transplant market.

the share of those new transplant patients receiving the new therapy, and the ABM can be simulated multiple times to produce a distribution of possible outcomes, as shown in Figure 9.

Because the ABM approach used in this example employs simulation agents, it is possible to have each specific patient represented by its own simulation entity. Additional patient attributes (such as gender, age, geographic location, comorbidities, etc.) would be easily added to this ABM framework for additional detail. With this ABM simulation methodology established, the product's Brand Planning team would be able to assess not only the possible range of outcomes based on variability of input assumptions, but also to dissect the resulting patient base along a large number of attribute dimensions. Given the small patient pool involved in this particular area (pancreas transplant), such investigations would likely be unfeasible with a traditional disaggregation approach.

Detailing and WOM in Doctor Adoption Framework

The ABM framework is increasingly used in dynamics involving physicians and the marketing activities designed to increase their adoption and use of pharmaceutical products. Chapter 8 covered the traditional Dynamic Modeling approach to dealing with this particular sector under the heading of Doctor Adoption. With a growing collection of historical data upon which to base so-called social network interactions, ABM provides the cutting-edge approach to analyzing physician adoption behavior.

For example, suppose that a Brand Planning team is designing a physician marketing strategy for a soon-to-be-released compound. Specialists are the

Table 2 Hypothetical Adoption and WOM Parameters as a Function of Detailing Efforts

Cumulative number of details (*per doctor*)	Probability of adopting (*in a given year*)	Likelihood of mentioning compound to colleague (*in a given year*)
0	0.01	0.00
1	0.05	0.02
2	0.11	0.10
3	0.18	0.15
4	0.22	0.18
5	0.25	0.20

targeted doctor group, as the indication for which the drug was developed is rarely if ever treated by Primary Care Physicians (PCPs) or general practitioners. For simplicity, we will assume that detailing on the part of the company's sales representatives is the only way to get doctors to try and ultimately adopt the product. In addition, conventional wisdom suggests there is a strong WOM effect within this specialist group.

For the sake of illustration, suppose a series of market research studies and analysis of longitudinal prescribing behavior related to previously released compounds indicates a very definitive relationship between detailing and the related adoption and WOM parameters, as shown in Table 2.

In other words, the data in Table 2 indicate that increasing the number of detailing visits increases not only the rate at which those targeted doctors will adopt the product, but also the likelihood they will mention the product to a colleague. Suppose further than the primary market research studies indicate that a WOM referral (from one doctor to another) is regarded as twice as valuable as a detail from a company sales representative. In effect, a doctor with no previous knowledge of the product would subsequently behave the same if he or she received two detailing visits *or* if he or she heard about the product from a colleague. Although sometimes difficult to quantify, these are exactly the types of social interaction dynamics for which ABM is an appropriate solution.

Imagine the indecision on the part of the Brand Planning team regarding their marketing strategy, given the complexity of this situation. How should the team most effectively allocate resources in order to maximize physician adoption of the compound? Would a broad strategy be warranted, executed by first detailing *all* of the targeted doctors once and then again and finally a third time and so on? Or would it be more effective to isolate a small group of specialists, detail them intensively over a short period, in the hopes that they would both adopt the product quickly and influence their colleagues to do the same via WOM?

This type of problem is fairly intractable using traditional analytic approaches, but an ABM construct is relatively easy to construct, implement, and analyze. First, an adoption framework using stocks and flows would be established.

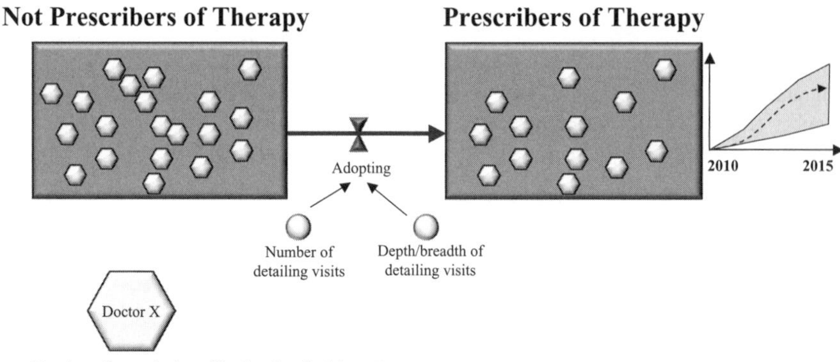

Figure 10 Example ABM Doctor Adoption framework incorporating detailing and WOM effects.

Next, this dynamic model would be populated with simulation agents representing all the doctors in the target population. These agents would be given characteristics or attributes related to the number of times they had been detailed and the number of times they had heard about the product from their peer group. The data in Table 2 would be used to govern the behavior of those agents/doctors as they interacted with sales reps and other specialists throughout the country. Finally, the simulation would be established to look at the effects of various physician marketing strategies, through which the Brand Planning team could examine the impact of both the number of detailing visits to doctors and how they were either focused or dispersed. Figure 10 shows how such an ABM might be established.

By looking at how the ABM responds to changes in the magnitude and distribution of detailing visits, the Brand Planning team can formulate a more effective allocation of marketing resources. In addition, the team will have a much better idea as to the expected trajectory of physician adoption of the product, which ultimately impacts the sales forecast for the compound. Often such ABMs are integrated with a Patent Flow and Treatment Attractiveness components of the basic Dynamic Modeling framework to result in an integrated model of marketplace dynamics. Some of our clients layer in geographic, prescription volume information, and attitudinal profiles of their targeted physician base into the ABM framework to attain an even more robust picture of the Doctor Adoption process.

TOOLS FOR CREATING AGENT-BASED MODELS

The past decade has seen a big increase in both the number and the quality of simulation packages designed to create ABMs as well as a corresponding jump in the number of business applications to which this methodology is being applied. Although this is a relatively new and sophisticated approach, we find ourselves fielding more questions about the ABM framework from clients and other interested parties. Potential ABM practitioners have a number of packages to choose from (such as AnyLogic®, Swarm, NetLogo, Repast, and Mason to name a few) and will need to evaluate the pros and cons of these tools in order to make an appropriate choice of software.

The vast majority of ABMs created within our consulting practice are constructed using the AnyLogic application, as we have found this package offers the best tradeoff of features and flexibility. AnyLogic has the following advantages:

1. An ability to combine ABM and traditional SD frameworks within the same model.
2. A framework that allows for a large number of simulation agents.
3. A relatively easy-to-use interface.
4. A method for tying model results to a visual schematic, making for very nice "real time" simulation outputs.
5. An efficient simulation algorithm so the resulting ABM executes relatively fast.

Other practitioners find similar benefits with some of the other modeling packages, and we have found that sound application of Dynamic Modeling principles to the ABM arena is more important than the choice of any particular software application. Any one of these programs will give the beginning modeler a very sophisticated tool with which to explore this exciting new field.

STRATEGY TESTS USING ABM FRAMEWORK

The Dynamic Modeling approach has typically had a foot in two different worlds—one of strategy development and one of forecasting/analytics. The inclusion of ABM principles to traditional dynamic models has tended to push the focus toward the latter of these two endeavors, simply due to the technical sophistication and computer simulation requirements of the ABM framework. However, advanced approaches such as ABM still have an important voice in the realm of strategic initiatives. In the pharmaceutical world, Brand Planning teams are often not as concerned about the details of a particular modeling approach, but rather that the tools employed are well suited to provide meaningful answers to specific problems. Chapter 12 will illustrate how the Dynamic Modeling framework can be used to ask and ultimately answer important strategic questions from a Brand Planning perspective, and the approaches outlined therein are valid regardless of the underlying model structure. In fact, we have found that the progression of strategy tests described in chapter 12 work well with a range of model architectures, from deterministic, traditional system dynamics frameworks all the way to probabilistic, ABM approaches. Readers with technical backgrounds might be more interested in the inner workings of simulation algorithms and how they can be applied analytically, while those with a strategic bent might be more concerned with how Dynamic Modeling can help answer some "big picture" questions. Brand Planning teams often want the perspective provided by both of these lenses, and the need to be both analytically rigorous and strategically relevant is well served by Dynamic Modeling in general and the ABM framework in particular.

SUMMARY

- Dynamic models in the pharmaceutical world are increasingly utilizing more sophisticated approaches to better leverage historical data regarding patient and physician behavior.
- ABM is a methodology in which simulation entities are created to represent individual items (often patients, doctors, or developmental compounds) in a dynamic system.
- ABMs have advantages in certain circumstances where traditional segmentation or disaggregation approaches become unwieldy.
- The use of agents in a simulation is simply another way to employ the Dynamic Modeling methodology, and can benefit from the same types of strategy tests used to analyze more traditional frameworks.
- There are a number of available software tools for creating, simulating, and analyzing ABMs.

REFERENCE

1. The Organ Procurement and Transplant Network, 2008. http://www.optn.org.

10

Treatment Attractiveness of Pharmaceutical Products

INTRODUCTION

This chapter will cover the following topics:

- Introduction to attractiveness/utility
- Assumptions behind attractiveness scoring
- Aggregate Treatment Attractiveness
- Factors of Treatment Attractiveness
- Varying Treatment Attractiveness by patient segment
- Options for reducing treatment options list

Recall in chapter 2 we discussed the three main sectors of the Standard Template. First, the patient flow component determines the stages of, and dynamics associated with, the patient movement and represents places in which potential patients can be captured. Second, the Doctor Adoption piece details the diffusion process through which newly released drugs will be going; as they are accepted by the base of prescribing physicians. Third, the area of Treatment Attractiveness covers the relative evaluation of the drug options in the marketplace. It is to this last sector that we now turn our attention.

To do so, we must depart from the operational "physics" of stock/flow structures to a slightly more calculation-based approach when determining the attractiveness of various pharmaceutical offerings in a given market.* When we have completed this task, we will bring the three sections together to show how they interrelate to form a fully functional Standard Template dynamic model.

* Interested readers are encouraged to investigate *Forecasting for the Pharmaceutical Industry: Models for New Product and In-Market Forecasting and How to Use Them* (1) for a comprehensive yet highly readable account of how such concepts are incorporated into other projection methodologies.

We all have had first hand experience with the concept of attractiveness* in our everyday lives. Ratings of different wines by sources such as Robert Parker or Wine Spectator abound at your local liquor store or wine boutique. These wines are rated on a 0–100 scale, with most widely available wines falling in the 80–95 range. Details on how these determinations are made are rarely made available, but the overall metric of utility has become an industry standard. Hotels and movies are often rated on a 5-star system by experts or critics in the field. Many consumer magazines such as *Consumer Reports* or *PC World* evaluate products based on a 0–100 scale, usually providing the detailed explanation of the means by which the overall metric of attractiveness was calculated.

In each of these cases, utility evaluations are designed to provide the consumer with a consistent, bounded scale on which various offerings can be compared. The ratings are intended as guides to attractiveness—the average consumer will derive more benefit out of a more highly rated product or service than a lower one. These evaluations are by no means "scientific"—in most cases, they represent the opinions of either a single person or small group of industry pundits. Nor are they universal or applicable to every consumer—a given individual may love a movie rated only 1 star or hate a wine given a blockbuster score of 98. Regardless, product attractiveness evaluations do provide a useful set of metrics that provide information to help consumers make product comparisons.

In the Standard Template, we bring this concept of utility to the pharmaceutical world by incorporating Treatment Attractiveness scores for various treatment options and calibrating them to actual market-level data. Incorporating specific choice algorithms (see chap. 10) with these pharmaceutical product utility scores results in an integrated dynamic model that can be used to derive expected patient flow dynamics throughout a forecast interval.

To complete this last piece of the Standard Template dynamic model, the overall utility of each treatment option needs to be determined on a consistent, bounded scale. We will describe a number of ways to achieve this endpoint, but bear in mind the details of the approach are not as important as the overall objective. For example, we often accept the overall evaluation of a film by a movie critic without knowing the precise method by which that ultimate rating was made. Similarly, the details through which the overall treatment attractiveness measure can be derived may seem tedious and overly complicated, but the end result is a simple set of utility numbers that relate to the products in our market definition.

As we relate these topics back to the pharmaceutical industry and the Standard Template dynamic model, a number of unique features become readily apparent, as follows:

- The consumer and the decision maker are different individuals, as it is the physician who evaluates and ultimately chooses between available products but the patient who abides by that decision and uses the chosen product.

* Economists prefer the term "utility" to "attractiveness," but we will use these terms interchangeably.

- Established utility scales are generally nonexistent for pharmaceutical products, and hence such attractiveness metrics must be derived and agreed upon rather than simply collected from published data or secondary market research.
- For the purposes of forecasting, utility scores of currently available products *and* their soon-to-be-released counterparts must be established. The attractiveness metrics for pipeline drugs are difficult to determine and are often revised as better clinical trial data becomes available.

Given that doctors in most cases make the decision regarding the appropriate prescription regimen for their patients, Treatment Attractiveness should be viewed through the eyes of prescribing physicians. Direct-to-consumer advertising efforts, prescription suggestions and/or changes on the part of nurse practitioners or pharmacists, and access to prescription drug information through the Internet are slowly influencing this paradigm, but physicians are still the primary player in the prescribing decision. In the sections that follow, all evaluations of treatment attractiveness will be viewed as applying to the doctors who treat patients in a given indication, not the patients themselves.

INITIAL ASSUMPTIONS INVOLVED IN TREATMENT ATTRACTIVENESS SCORING

We can begin to tackle the utility evaluation process by making two key initial simplifications:

1. *Nondifferentiated physician universe*—all doctors are the same and will have identical evaluations of the pros and cons of various treatment options in the marketplace.
2. *Product knowledge is universal and infallible*—all doctors know everything there is to know about each therapy and that knowledge is not subject to interpretation or influence.

These assumptions will be relaxed as the Standard Template model is extended, but provide a useful and consistent initial starting point for analysis.

This process of determining the individual Treatment Attractiveness scores for each element of a mutually exclusive and collectively exhaustive (MECE) treatment option set has one goal in mind—**an evaluation of utility for each therapy regimen on a consistent, bounded scale**. These scores are then used as part of an algorithm that determines the allocation of the flows of patients in the Standard Template.

To begin, recall our continuing example of a market for a disease that has five therapy regimens: Treatments A, C, D, BC, and other. Treatment D is currently not available, but launches in early 2013. Prescribing physicians in this market likely have a perhaps implicit evaluation of how these drug regimens rank relative to one another. In theory, these evaluations provide the basis for the decision to

Table 1 Example for Aggregate Treatment Attractiveness Scores (2010–2015)

Treatment option	2010	2011	2012	2013	2014	2015
A	45	45	45	45	45	45
C	50	50	50	50	50	50
D	0	0	0	55	55	55
BC	48	48	48	48	48	50
Other	50	50	50	50	50	50

prescribe one therapy instead of another. The Treatment Attractiveness section of the Standard Template allows us to explicitly capture this evaluation process.

DETERMINING AGGREGATE TREATMENT ATTRACTIVENESS

One approach to evaluating Treatment Attractiveness is to have members of a cross-functional team simply score these five drug regimens on some scale—typically 0–100. The upper bound of 100 represents the best or optimum drug imaginable, existing or otherwise. The lower bound of 0 represents the worst drug possible—sometimes thought of as the score a placebo would receive. The midrange value of 50 indicates an average drug for the indication. In our experience with clients, the process of finding a consensus estimate of relative treatment utility often results in a reasonably accurate assessment. The results of such an evaluation process are shown in Table 1.*

The example in Table 1 shows Treatment Attractiveness scores that stay constant over time, but incorporating changing utility metrics is often done as a result of a team's evaluation process.

This aggregate assessment of attractiveness provides a useful set of inputs to the Standard Template. However, some teams choose to expand this evaluation by addressing the specific factors that ultimately determine such metrics. Comparing the drugs in the model's MECE set according to specific product attributes allows for a more detailed evaluation of Treatment Attractiveness to take place.

DETERMINING FACTORS OF TREATMENT ATTRACTIVENESS

A number of factors can impact the aggregate utility calculation for treatment options in a given indication, which can be imagined as a "roll up" of the treatment attributes that physicians consider important in evaluating each treatment option. The key attributes in the decision making process vary by indication, but often include:

- Safety
- Efficacy
- Side effects

* Note the utility score of Treatment D is 0 until it launches in 2013.

Table 2 Example for Set of Key Product Attributes Determining Treatment Attractiveness

Treatment attributes
Efficacy
Safety
Onset of action
Side effects

- Tolerability
- Mode of administration
- Onset of action

To establish a more detailed evaluation of Treatment Attractiveness, a Brand Planning team must establish the important components of therapy utility in a particular disease market. For example, input from a team's clinical and/or medical functions, coupled with primary or secondary market research, may indicate 3–4 key attributes that have a significant impact on a physician's prescribing decision, as shown in Table 2.

From a cause-and-effect standpoint, including these treatment attributes shows that each plays a major role in determining Treatment Attractiveness for each therapy regimen in an MECE set, as shown in Figure 1.

DETERMINING IMPORTANCE OF TREATMENT ATTRACTIVENESS FACTORS

With the set of causal factors established, the Brand Planning team can determine the importance of each factor in determining aggregate Treatment Attractiveness. For example, a treatment regimen's safety profile may be a hugely important issue in a disease market if drug toxicity problems plague the current treatment landscape. In such cases, the safety attribute may constitute a high percentage of the overall Treatment Attractiveness calculation. In other markets where there are well established and comparable products in terms of efficacy, the side effect attribute may be of primary importance. Assigning weights to each treatment attribute will establish what percentage each factor plays in the aggregate evaluation of utility, as shown in Table 3.

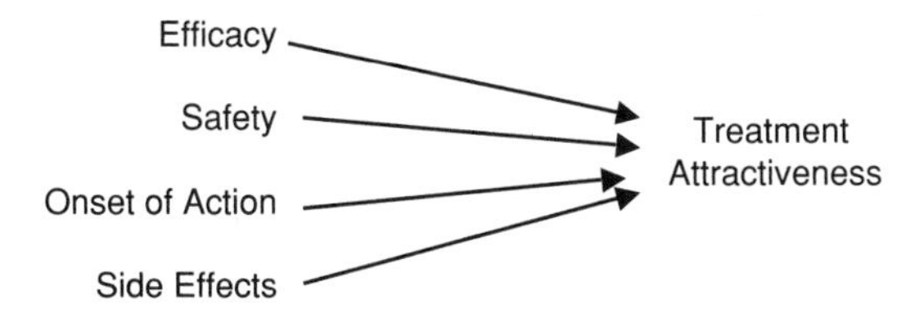

Figure 1 Causal factors determining aggregate Treatment Attractiveness.

Table 3 Example for Treatment Attribute Weightings

Treatment attribute	Weight–Year 2010 (%)
Efficacy	30
Safety	30
Onset of action	25
Side effects	15
Total	100

There is no single "right" answer regarding the importance of key treatment attributes such as those shown in Table 3. Published market research can often give some indication as to the relative magnitudes of these types of importance evaluations, and some pharmaceutical firms conduct extensive physician surveys to arrive at these estimates. From a Brand Planning perspective, however, it is vital that these estimates are agreed upon by various members of a cross-functional team. Building consensus around the components of Treatment Attractiveness establishes buy in and helps to ensure that overall Dynamic Modeling results receive proper consideration.

Strategic Implications of Treatment Attractiveness in Dynamic Modeling

The process of establishing the treatment attractiveness metrics for both current and pipeline products in a given indication is an extremely valuable part of effective Brand Planning. Without the rigor of utility formulations, various functions within an organization often operate with implicit assumptions about how individual treatments stack up relative to one another, how soon-to-be released drugs would fit into the overall treatment landscape, and on what basis physicians evaluate the set of therapy options in a given disease marketplace. By requiring these assumptions to be quantified in the Treatment Attractiveness framework, input from members of the Brand Planning team is clearly spelled out and integrated in a systematic way. The scoring process helps teams share information sources, build agreement across functions, and achieve consensus regarding the relative utility of drugs in the current treatment landscape and how these metrics might evolve over time with the introduction of new compounds.

Treatment Attractiveness scores are useful in isolation, but are even more insightful when integrated into the Dynamic Modeling framework. Calibrating the model to historical data helps to validate the assumptions behind these utility metrics and ensures that the model is replicating actual market behavior. The validation stage is crucial to establishing buy-in from the Brand Planning team and ultimately communicating the insights of the Dynamic Modeling process throughout the organization.

Including the Treatment Attractiveness scores into the fully functional dynamic model also allows teams to do scenario testing with respect to the components of therapy utility. Sensitivity analysis, for example, can determine a range of possible outcomes for a particular compound based on different Treatment Attractiveness scores. Components of attractiveness can be analyzed as well, testing their impact on the future of a treatment in a disease marketplace. Finally, the importance of various clinical factors in physicians' algorithm for determining utility can be redistributed, indicating how much the rating of various attributes affects overall Treatment Attractiveness scores and subsequent market behavior.

Using dynamic models and their associated utility metrics provide Brand Planning teams with unique strategic insights into the determinants of marketplace dynamics. The process of explicitly formulating Treatment Attractiveness, combined with the ability to test the strength of the components of that formulation, enables a better understanding of the markets in which individual compounds might play. And simulation can help Brand Planning teams to better focus their efforts on the areas of Treatment Attractiveness that have the most strategic leverage.

DETERMINING IF IMPORTANCE OF TREATMENT ATTRACTIVENESS FACTORS CHANGES OVER TIME

Dynamic models are designed to capture behavior changes over time, so once the list of treatment attributes is identified and the associated weights are assigned, the working team can determine whether these weights will stay constant. Depending on the types of New Chemical Entities (NCEs) that may be launched in the years ahead, or the macroeconomic factors influencing the evolution of health care, the competitive landscape could change significantly in the future. Allowing for the importance of treatment attributes to change over time provides the means for the Standard Template model to replicate a dynamic competitive marketplace. In our continuing example, changing weights can be assigned to the treatments in the MECE set for each year from 2010–2015, as seen in Table 4.

Note that these treatment attribute weights do not have to change over time, but their variable nature is easily incorporated using Dynamic Modeling simulation technology. The design of the approach gives it the flexibility to handle variable treatment attribute weightings just as easily as constant weightings.

DETERMINING RATINGS OF TREATMENT ATTRACTIVENESS FACTORS

Once the treatment attribute weights have been assigned, the Brand Planning team can assign relative rankings by treatment attribute for each therapy in the MECE

Table 4 Example for Treatment Attribute Weightings Over Time (2010–2015)

Treatment attribute	2010 (%)	2011 (%)	2012 (%)	2013 (%)	2014 (%)	2015 (%)
Efficacy	30	30	30	30	30	30
Safety	30	30	30	30	30	30
Onset of action	25	25	25	25	25	25
Side effects	15	15	15	15	15	15
Total	100	100	100	100	100	100

set. Data from a variety of sources help in this process. In our experience, team members from marketing, market research, clinical, medical, and even forecasting have access to different data sources and perspectives that can help provide a much more comprehensive evaluation of available drugs. For example, market share information may provide a rough estimation of how products are being accepted in the marketplace. Clinical trial data often provides valuable product comparisons across dimensions of attractiveness such as efficacy or side effects. And primary market research may have data on the strengths and weaknesses of drug options in the market. Incorporating this knowledge from each function allows for a more accurate view of the treatment landscape to emerge.

The team can now rank treatment attributes on a scale of 0–100, similar to the evaluation of overall utility seen earlier in this chapter. For example, Efficacy might be scored as shown in Table 5.*

The other dimensions of Treatment Attractiveness (Safety, Onset of Action, and Side Effects) would be scored in a similar fashion.

CALCULATION OF TREATMENT ATTRACTIVENESS

In the manner described above, overall Treatment Attractiveness is disaggregated into separate rankings for component factors. Instead of an agreed-upon aggregate Treatment Attractiveness score, utility now becomes a weighted average of various

Table 5 Example for Efficacy Ratings Over Time (2010–2015)

Treatment option	2010	2011	2012	2013	2014	2015
A	50	50	50	50	50	50
C	60	60	60	60	60	60
D	0	0	0	65	65	65
BC	55	55	55	55	55	55
Other	56	56	56	48	45	40

* Again, Treatment D scores a "0" on this dimension until launch.

Table 6 Example for Calculation of Treatment Attractiveness—Treatment A (2010)

Treatment A–Year 2010			
Attribute	Score	Weight (%)	Treatment attractiveness from:
Efficacy	50	3	15
Onset of action	50	30	15
Safety	10	25	2.5
Side effects	50	15	7.5
Total		100	40

attractiveness attributes. For example, the calculation for Treatment A's utility in 2010 is shown in Table 6.

The calculation shown in Table 6 is performed for individual drugs in each year of the forecast interval. The resulting calculations from the model structure and the equations functioning within this framework create dynamic Treatment Attractiveness scores over time, as shown in Table 7.

Again, these utility evaluations assume that *all* prescribing doctors are aware of and willing to prescribe the five treatments options and have complete knowledge of all regimens in the consideration set. This assumption will be relaxed in chapter 11 when Treatment Attractiveness will be combined with the patient flow and Doctor Adoption sectors to arrive at the fully integrated Standard Template dynamic model.

TREATMENT ATTRACTIVENESS BY PATIENT SEGMENTATION

As discussed in chapter 5, the Standard Template allows for disaggregation of patients into categories such as gender, age, and/or severity of disease. Creating more detailed patient categories would provide an avenue for an extension of the

Table 7 Example for Calculated Treatment Attractiveness Scores Over Time (2010–2015)

Treatment Option	2010	2011	2012	2013	2014	2015
A	40	40	40	40	40	40
C	53	53	53	53	53	53
D	0	0	0	52	52	52
BC	52	52	52	52	52	52
Other	50	50	50	50	50	50

Table 8 Example for Treatment Attractiveness Scores by Disease Severity (2010)

Treatment	Mild	Moderate	Severe
A	40	40	55
C	20	50	80
D	0	0	0
BC	48	48	48
Other	80	60	10

Treatment Attractiveness evaluation, as utility scores can be specifically matched to individual patient segments and/or simulation agents.*

For example, the cross-functional team may want to understand the size of the patient pools and the corresponding treatment landscape across dimensions of disease severity. Identifying these differences may be important for understanding the treatment dynamics that ultimately drive strategic focus. Including this patient segmentation opens the possibility for each drug's utility score to vary for mild, moderate, and severe patients. An example of such an evaluation of Treatment Attractiveness by severity is shown in Table 8.

As shown earlier in this chapter, the analysis of Treatment Attractiveness by disease severity could be extended through each year of the forecast interval. Additionally, the aggregate utility metric for each treatment option could be factored down to include an evaluation of the detailed treatment attributes contributing to it.

Matching Treatment Attractiveness scores with the corresponding dimensions of a patient disaggregation allows for a finer degree of analysis for detailed patient groupings. If data is collected on how treatment patterns and treatment dynamics vary by patient severity, for example, these data can be calibrated to severity-specific utility scores in the Standard Template model. Patient segmentation and corresponding utility evaluations often highlight important differences in an indication marketplace—differences which can have huge impact on product positioning and strategic focus.

STEPS IN THE TREATMENT ATTRACTIVENESS PROCESS

Our clients have found it useful to think about the Treatment Attractiveness scoring exercise using the chart shown in Figure 2. At each step, the Brand Planning team must determine the value in progressing to the next level of detail in determining therapy utility in a given indication marketplace. Again, there is no single right

* The process can also be applied to physicians such that the utility scores for therapy options vary according to specialty. This is especially useful in markets where PCPs have vastly different prescribing patterns than indications specialists, even for identical patient types.

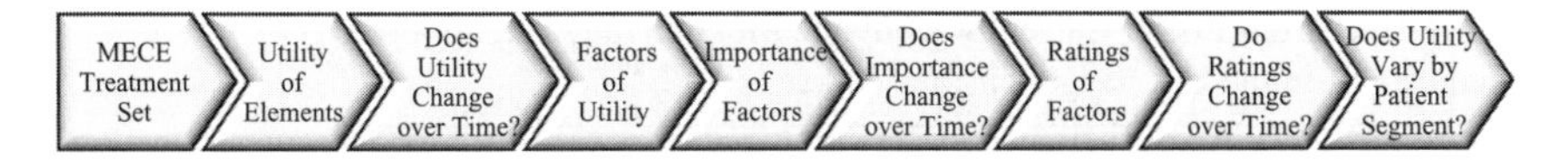

Figure 2 Decision chart in determining Treatment Attractiveness scores.

answer regarding this process—the requirements for each project will determine a suitable level of Treatment Attractiveness detail.

EXTENSION: REDUCING THE NUMBER OF TREATMENT OPTIONS IN THE MECE SET

The example detailed in this chapter deals with a small set of only five treatment options. What happens when a larger number of drug regimens exist in a particular market? The approach we have outlined could certainly be extended ad infinitum, but as a practical matter, dynamic models work best with no more than 15 or so options in the MECE treatment set. Instead of extending the number of regimens included in the Standard Template dynamic model beyond this limit, we have found two useful alternatives:

1. Group similar drugs into a default category that becomes a unique element in the MECE set.
2. Use a two-tiered approach in which a drug class is chosen first and then drugs within that class are evaluated.

Instead of addressing these options in abstract terms, we will look at the monotherapy hypertension market to see how these two alternatives could be applied. Table 9 shows list of some of the treatment options available in this market. Although this list is not comprehensive and does not address possible combination usage, it will provide a useful example of potential approaches to limiting the number of therapy options to a manageable level.

Option 1 for Managing Treatment Option Complexity: Group like drugs into a default category that becomes a unique element in the MECE set

If the working team is most concerned with analyzing large, name brand treatment options, then grouping smaller products into a collective element in the MECE is often productive. This approach is useful when analyzing the individual behavior of *each* drug in a marketplace is unnecessary. This collective grouping becomes a de facto treatment option by itself, and the team can score this group of drugs as an individual element as opposed to doing so for each drug individually.

The list of hypertension monotherapy options in Table 9 has 14 possibilities—suggesting some sort of grouping is necessary to reduce the treatment option set. The goals of the Brand Planning team will dictate the correct procedure for collapsing treatment options into a default category. In this example, the team's interest in analyzing strong branded competitors suggests the remaining

Table 9 Partial List of Currently Available Monotherapy Prescription Medications in the U.S. Hypertension Market

Antihypertensive class	Chemical	Tradename
ACE inhibitors	Ramipril	Altace
	Lisinopril	Prinivil/Zestril
	Enalapril	Various
	Other ACEs	Various
ARBs	Irbesartan	Avapro
	Losartan	Cozaar
	Valsartan	Diovan
	Candesartan	Atacand
	Other ARBs	Various
Calcium channel blockers	Amlodipine	Norvasc
	DHP	Various
	Non-DHP	Various
Beta blockers	All BBs	Various
Diuretics	All diuretics	Various

drugs be grouped into one category "catchall" category called "other," as shown in Table 10.

By collapsing various treatment options into a catchall category, the MECE set has been reduced to eight—a number easily managed from a data and ease-of-use standpoint. The elements in this revised treatment set can then be subjected to a Treatment Attractiveness scoring exercise as discussed earlier in this chapter.

Calibration to patient-level data can be tricky when grouping a large number of drugs into a catchall category. In some cases, a weighting factor is applied to the collective category to capture the number of treatment options contained in it. This additional factor should be added with care, however, to ensure that the

Table 10 Collapsed List of Currently Available Monotherapy Prescription Medications in the U.S. Hypertension Market

Antihypertensive class	Chemical	Tradename
ACE inhibitors	Ramipril	Altace
	Lisinopril	Prinivil/Zestril
ARBs	Irbesartan	Avapro
	Losartan	Cozaar
	Valsartan	Diovan
	Candesartan	Atacand
Calcium channel blockers	Amlodipine	Norvasc
Other	Various	Various

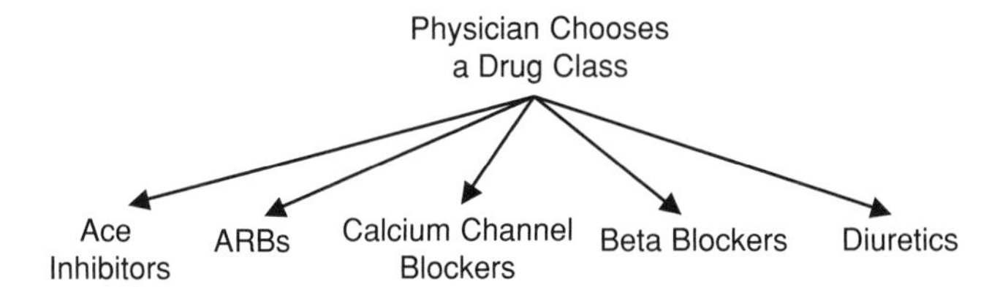

Figure 3 Choosing class first in a hypertension two-tier choice model.

Treatment Attractiveness calculation is not simply force-fitted to calibrate with actual historical data.

Option 2 for Managing Treatment Option Complexity: Use a two-tier approach in which a drug class is chosen first and then drugs within that class are evaluated

In some markets, physicians may first make a decision regarding which class of drugs to prescribe and then subsequently decide which drug to prescribe within that class. This process is defined as a two-tier choice model, in which individual drug selection takes place *after* a class choice is made. One advantage of this situation is that it can narrow the list of possibilities in the MECE set included in the model.

If the hypertension market shows signs of being a two-tier choice model, a prescribing physician must first choose which class of drug to prescribe: angiotensin-converting enzyme (ACE) inhibitors, angiotensin receptor binders (ARBs), calcium channel blockers (CCBs), beta blockers, or diuretics. This choice process is depicted in Figure 3.

To render this class choice as part of the Standard Template dynamic model, the cross-functional team can determine Class Attractiveness scores for the listed drug classes. It is important to note that in this framework, class allocation does *not* depend on the number of drugs within that class.

The second step in a two-tier choice algorithm is the choice of drug option within a particular class. Continuing the hypertension example, suppose that the ACE inhibitor class was chosen in the first step of the process. The subsequent decision involves choosing the particular drug within that class, as depicted in Figure 4.

Relative Treatment Attractiveness scores will determine how individual drugs are chosen within a given class; the scoring exercise to determine these therapy utilities mirrors that described earlier in the chapter.

By establishing a multistep process for drug selection, a two-tiered choice model in fact relies on separate attractiveness evaluations. First, classes of drugs are compared based on their Class Attractiveness metrics and a single class is chosen. Then, drugs within that class are evaluated according to their Treatment Attractiveness scores and a single drug out of the class set is chosen.

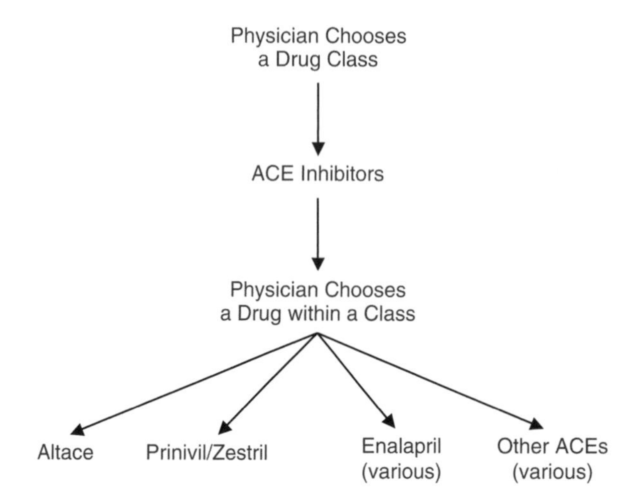

Figure 4 Choosing drugs within a class in a hypertension two-tier choice model.

Collapsing a treatment set is an important option to have within the Dynamic Modeling process, and can often result in an end product that is more manageable from both a data collection and model calibration perspective. The reduction method chosen will depend on a number of factors—goals of the modeling process, degree of substitutability between therapy options, data availability, etc. However, the methods shown in this chapter have the advantage of utilizing the attractiveness/utility approach and then seamlessly incorporating into the Standard Template framework.

Chapter 11 will bring the concepts of patient flow, physician adoption of product, and Treatment Attractiveness/utility together in a simulation framework to show how, together, these three sectors can be used to create a "virtual marketplace."

SUMMARY

- Attractiveness/utility is a common metric in the world of consumer goods.
- Applying the concept to pharmaceutical products requires a rigorous approach.
- Aggregate utility scores can be applied to all the therapy options in a model's MECE set, and can be designed to change over time.
- Alternatively, factors contributing to Treatment Attractiveness can be assigned by the working team.
- The importance and rating of each of these treatment attributes can be determined by team judgment for each of the MECE treatment set, and can be designed to change over time.
- The Standard Template can calculate changing Treatment Attractiveness scores, based on assumptions regarding treatment attributes.

- Utility calculations can be established specific to patient segmentation.
- In markets with large numbers of treatment options, the MECE set can be reduced by collapsing similar drugs into a default category, or by establishing a two-tier choice model with corresponding utility metrics at both the class and product level.

REFERENCE

1. Cook, Arthur G. Forecasting for the Pharmaceutical Industry: Models for New Product and In-Market Forecasting and How to Use Them. Aldershot, Hampshire, England: Gower Publishing, 2006.

11

Integrating the Three Sectors of the Standard Template Dynamic Model

INTRODUCTION

This chapter will cover the following topics:

- Logit Choice algorithms
- Arithmetic Logit Choice (ALC) formulation
- Exponential Logit Choice (ELC) formulation
- Applying choice algorithms to patient flow dynamics
- Incorporating doctor adoption/prescribing dynamics
- Full Standard Template diagram

Even in isolation, the three sectors of the Standard Template provide insight into the driving factors in pharmaceutical marketplaces. But the true power of the Dynamic Modeling approach is found when these sectors come together to form a fully integrated, operational representation of the drivers of change in various disease indications. This chapter will introduce the integration of Patient Flow, Doctor adoption, and Treatment Attractiveness structures in a series of steps designed to walk the reader through the calculations and methodology involved in bringing these three key pieces together in a coherent Dynamic Modeling framework.

LOGIT CHOICE ALGORITHMS

Chapter 10 in our continuing example demonstrated how a numerical evaluation of Treatment Attractiveness for elements in a mutually exclusive and collectively exhaustive (MECE) treatment option set can be established. With this evaluation of utility established, we can now incorporate the vast body of work in the area of choice formulations. A word of encouragement as we proceed: many interested parties are scared off by the equations and mathematical lingo that usually accompanies such choice algorithms. The concepts behind these approaches are actually quite simple, however, and can be described using a combination of numerical and verbal depictions that makes them readily accessible.

Recall that the process of determining the individual Treatment Attractiveness scores for each element of an MECE treatment option set has one goal in mind; *an evaluation of utility for each element on a consistent, bounded scale.* These scores are then used as part of an algorithm that determines the allocation of the flows of patients in the Standard Template.

Combining Aspects of Patient Flow, Doctor Adoption, and Treatment Attractiveness

Suppose a Brand Planning team at a pharmaceutical firm has established an MECE list of treatment options and their associated attractiveness scores for a particular indication, as shown in Table 1.

Note that Treatment D is being launched in 2013, and hence has a utility score of 0 in the years before it is commercially available. In effect, this means the competitive set for 2010–2012 consists of four treatment options: A, C, BC, and "other."

To proceed with this analysis, we will begin with two simplifying assumptions regarding prescribing physicians in this example marketplace:

1. *All prescribing doctors have all four on-market drug options in their consideration set.* Every physician actively prescribes each treatment to at least some of patients, and will do so in the foreseeable future.
2. *Treatment D will have 100% doctor adoption at the time of launch and at all points in the future.* The launch of treatment D triggers an immediate entry into physicians' consideration set, putting it on par with the four existing therapy options, at least in terms of doctor acceptance.

These assumptions will be relaxed as we progress through this chapter, but their inclusion at this point allows us to look at the basic calculations involved in the Logit Choice algorithm and later introduce the complexities of the Doctor Adoption component. Recall that the Treatment Attractiveness scores in Table 1 reflect market-level evaluation of the treatment options on an aggregate basis. To

Table 1 Example MECE Treatment Attractiveness Scores (2010–2015)

Treatment	2010	2011	2012	2013	2014	2015
A	45	45	50	50	50	50
C	60	60	55	50	50	50
D	0	0	0	50	60	50
BC	45	55	65	60	50	50
Other	50	50	50	50	50	50
Total	200	210	220	260	260	250

Table 2 Example Treatment Allocation for Newly Diagnosed Patients (2010–2011)

Treatment	2010 (%)	2011 (%)
A	22	20
C	30	30
D	0	0
BC	20	28
Other	28	22
Total	100	100

fully utilize them, we must first attempt to relate these utility scores to actual market data.

Suppose a primary market research study obtained data on the treatment initiation paradigm of newly diagnosed patients. To return to the stock/flow structure of the Standard Template, this corresponds to the allocation of the flow of *diagnosed incidence* across the various categories of treatment options in the MECE set. The research showed a distribution of first-time treatments for newly diagnosed individuals, as shown in Table 2.

Conceptually, we now have an evaluation of treatment option attractiveness/utility (Table 1) *and* market-level data (Table 2) to which it can be compared. A mathematical algorithm can then be used to establish a numerical relationship between these two datasets.

Even before the introduction of statistical choice models to analyze these data, however, a few aspects of the form and implication of these data should be readily apparent, as follows:

- The ordering of the utility scores maps very well to the ordering of the share scores. Treatment C has the highest score and the highest observed share, while Treatment A has the lowest.
- The MECE set definition ensures the total shares of the treatment options = 100%.
- Changes in observed shares do not automatically corresponded to changes in utility scores. The Treatment Attractiveness score for Treatment other, for example, stays constant at 50 but its corresponding share drops from 28% to 22%.
- Share numbers tend to cluster around the 25% level, and the corresponding Treatment Attractiveness scores tend to center around the midpoint score of 50.*

The marketing literature suggests myriad choice models that can be applied in various situations (1). Our client teams generally use two types of fairly basic algorithms to calibrate scores to shares with these types of data.

* This observation is not true in every case, only of note in this specific example, as in many real-world markets the distribution of utility scores and share numbers varies greatly.

Table 3 Example ALC Calculations and Derived Patient Share Results (2010)

Treatment	Calculation	Derived patient shares (%)
A	45/200	23
C	60/200	30
D	0/200	0
BC	45/200	23
Other	50/200	25
Total		100

- ALC
- ELC

The ALC is the simplest algorithm to conceptualize and implement, and will be used as we extend this example. For readers with strong technical backgrounds (or those who simply love mathematical equations), the ALC formulation is:

Equation 10.1: ALC Formulation

$$S_n = U_n / \left(\sum U_x\right) + E_n$$

where S_n = the share for treatment n; U_n = the utility/Treatment Attractiveness score for treatment n; $\sum U_x$ = the total of all utility scores in the MECE set (x = 1 to n); E_n = the error term for the calculation, indicated the difference between the share calculation and the actual data to which it is being calibrated.

In layman's terms, this equation translates to "The calculated share for any treatment option is equal to the utility score for the treatment divided by the total of all utility scores for all the treatment options." So, for the first time period (2010 in this case), the ALC algorithm produces the share results according to a specific calculation, as shown in Table 3.

Putting the ALC in a Dynamic Modeling context allows this calculation to be made in any given time frame, so applying the algorithm to utility scores in each time period gives results shown in Table 4.

These outputs from the choice model can now be compared to actual historical data to determine the validity of the model approach. Various measures of "goodness of fit" exist for such calibrations, but even a cursory look shows this approach fits quite nicely in our example, as shown in Figure 1.

It is important to note that the application of the ALC or any type of choice algorithms will never be 100% "correct." All models, by definition, are abstractions from reality and hence "wrong" from the beginning.* Instead of focusing on complete accuracy, we are trying to find a framework to reasonably

* See chapter 2.

Table 4 Example ALC-Derived Patient Flow Share Results (2010–2015)

Treatment	2010 (%)	2011 (%)	2012 (%)	2013 (%)	2014 (%)	2015 (%)
A	23	21	23	19	19	20
C	30	29	25	19	19	20
D	0	0	0	19	23	20
BC	23	26	30	23	19	20
Other	25	24	23	19	19	20
Total	100	100	100	100	100	100

approximate the process and corresponding data we are attempting to emulate, usually for the purposes of better decision making. Even without the scrutiny of statistical analysis, the application of the ALC in this case provides results that are reasonably consistent with historical market-level data. Details of goodness of fit and statistical validity are beyond the scope of this work, but these analyses can easily be applied if a more rigorous examination of the ALC algorithm results is necessary.

We can now return to the stock/flow structure of the Standard Template for a moment to see how these choice models fit into the dynamic framework. Recall we are discussing the allocation of the flow of newly diagnosed patients as they enter the stock of Currently Treated Patients. Assuming universal and constant Prescribers status of 100% for each treatment option, as well as an ALC algorithm, the Dynamic Modeling diagram begins to take shape, as shown in Figure 2.

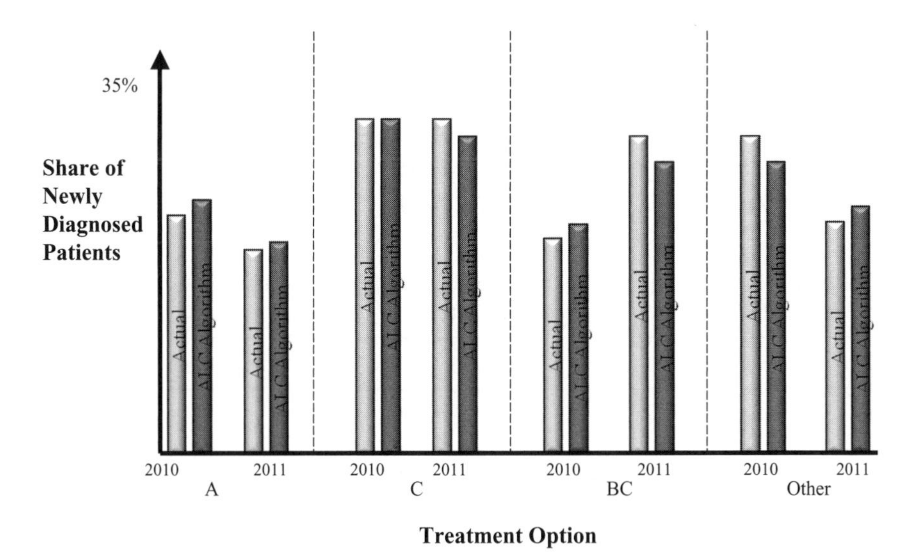

Figure 1 Comparison of example historical data to ALC-generated results.

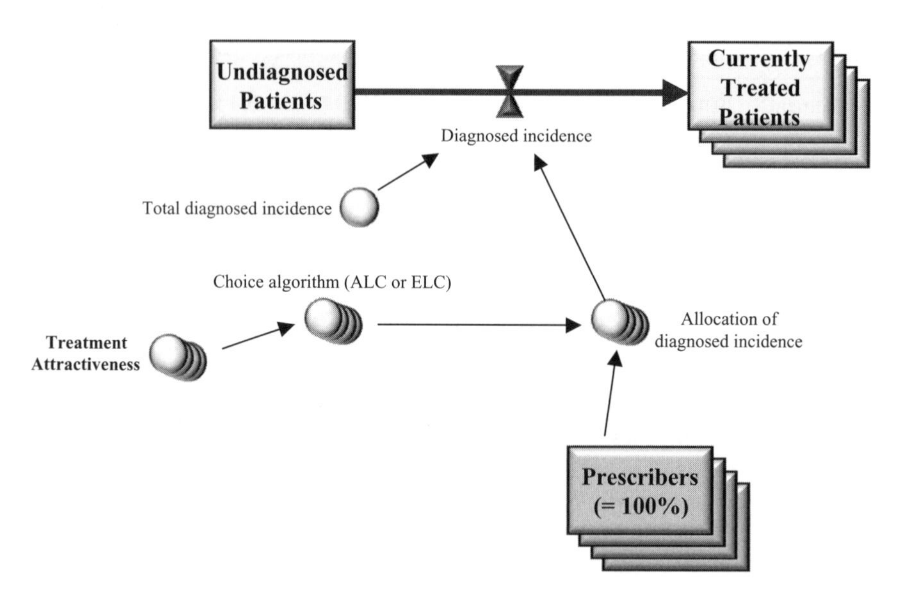

Figure 2 Introduction of the ALC algorithm to the Standard Template.

Recall from chapter 3 that the *total diagnosed incidence* flow in our continuing example is 16 and assume that the value stays constant over the forecast interval. The results of the corresponding dynamic model will take the form shown in Table 5.

EXTENDING TO OTHER PATIENT FLOW ALLOCATIONS

The process of application of a choice model to the stock/flow Standard Template can now be extended to other flows of interest; patients *reinitiating treatment* and *switching treatments.* In each, we follow the same process of using Treatment Attractiveness scores and finding an appropriate choice model that calibrates well

Table 5 Results of Applying ALC Formulation to Example Patient Flow of Diagnosed Incidence in the Standard Template (2010–2015)

Diagnosed Incidence	2010	2011	2012	2013	2014	2015
A	3.6	3.4	3.0	2.9	3.1	3.2
C	4.8	4.6	4.0	3.1	3.1	3.2
D	0.0	0.0	0.0	3.7	3.7	3.2
BC	3.6	4.2	4.7	3.4	3.1	3.2
Other	4.0	3.8	3.6	2.9	3.1	3.2
Total	16.0	16.0	16.0	16.0	16.0	16.0

to historical data. We will then show how to layer in the results of the Doctor Adoption structure.

Reinitiating Treatment

We can now turn our attention to the flow of patients who are *reinitiating treatment*, using the same approach of choice algorithms but now applying it to another flow in the stock/flow structure of the Standard Template. To follow the same procedure as we applied to *diagnosed incidence*, we first must gather information on the magnitude and allocation of the *reinitiating treatment* flow. Suppose we know from longitudinal analysis of patient-level data (see chap. 5) that approximately 25 patients return to treatment each year after being Non-Persistent. Additional market research may show these patients are allocated across our MECE set in the following distribution, as shown in Table 6.

We have already seen earlier in this chapter that the application of the ALC (using the Treatment Attractiveness scores as per our continuing example) produces share numbers that fail to show this degree of variation shown in Table 6, and are instead all in the 20% to 30% range. Application of the ALC algorithm in this case may produce results that are *not* reasonably valid in explaining the distribution of patients who are *reinitiating treatment*.

At this point the validity of the formulation (when applied to this specific situation) should be called into question, given the obvious disparities between ALC outputs and the data to which it is being compared. Such a situation often arises when using choice algorithms, and the decision options are fairly straightforward, as follows:

- Revisit the Treatment Attractiveness scores to arrive at utilities that better calibrate to historical data using the ALC. The trade-off in this specific case would be that utility scores with greater differentiation may no longer easily translate to the "clustered" distribution surrounding *diagnosed incidence*.
- Enact a different choice algorithm.

Table 6 Example Treatment Allocation for Patients Reinitiating Treatment (2010–2011)

Allocation of Reinitiating Patients	2010 (%)	2011 (%)
A	10	10
C	65	50
D	0	0
BC	10	25
Other	15	15
Total	100	100

Reevaluating utilities is a common occurrence in the Dynamic Modeling process, regardless of model validity or selected choice algorithm. In fact, teams will often rescore various treatment options as new data or information on changed perceptions become available. However, if a team is reasonably comfortable with the aggregate utility scores, a different choice formulation is often investigated in cases where the application of the ALC fails to provide a reasonable level of accuracy.

An ELC algorithm is extremely useful when utilities are clustered but market-level data shows wide disparities. The mathematics of the ELC are a bit more detailed, but the concept behind them is similar to the ALC formulation. In effect, the formula uses exponents to accentuate the differences in utility scores as they are translated into derived share numbers. Readers having bad flashbacks to college calculus courses might want to skip ahead a bit, but for the math enthusiasts out there, the formula for the ELC takes the following form:

Equation 10.2: ELC Formulation

$$S_n = e^\wedge (X^*U_n)/\left[\sum e^\wedge (X^*U_n)\right] + E_n$$

where S_n = the share for treatment n; e = Euler's number (~2.718); x = a constant to be solved for; U_n = the utility/Treatment Attractiveness score for treatment n; $\sum e^\wedge(x^*U_n)$ = the total of all utility scores when used as a power of e multiplied by some constant (x); E_n = the error term for the calculation, indicated the difference between the share calculation and the actual data to which it is being calibrated.

The mathematics of Equation 10.2 seems to be tricky, but are perhaps more easily conceptualized using a simple example. For any two numbers (say, 3 and 5) the difference between the powers of each number increases as the power increases, as shown in the following Table 7.

In similar fashion, the ELC uses exponents to magnify the differences between utility numbers. In this case, the variable x in Equation 10.2 refers to how much magnification is applied (i.e., to what power Euler's number is raised.) A small x means that differences in utility scores will not result in vastly different share calculations. A larger x implies small differences in utility scores can result in large differences in resulting share outputs.

Table 7 Differences Between the Powers of Two Example Numbers

Value	^1	^2	^3	^10
3	3	9	27	59,049
5	5	25	125	9,765,625
Difference	2	16	98	9,706,576

Table 8 Example ELC-Derived Patient Flow Share Results (2010–2015)

Dynamic Allocation ELC	2010 (%)	2011 (%)	2012 (%)	2013 (%)	2014 (%)	2015 (%)
A	9	7	9	13	13	20
C	64	52	18	13	13	20
D	0	0	0	13	48	20
BC	9	27	64	48	13	20
Other	17	14	9	13	13	20
Total	100	100	100	100	100	100

Statistical software packages or spreadsheet add-ons can be used to solve for x in order to best calibrate the ELC formulation to historical data. Certain Dynamic Modeling software platforms, such as VenSim® or AnyLogic®, have this capability as well. Alternatively, the ELC can be entered into Excel and x can be determined by trial-and-error or by using Excel built-in functions. Regardless of solution method, the ELC approach in this example case finds x to be approximately 0.13, and generates the share outputs as shown in Table 8.

Note the computed values in 2010 and 2011 compare nicely to the data collected in our hypothetical market research study (see Table 6).

Applying the ELC-derived patient flow allocation percentages to the flow of *total reinitiating treatment,* assuming this flow of patients stays constant at 25 per year, produces results shown in Table 9.

Switching Treatment

The dynamics of the *switching treatment* flow are more complex as they involve a 2-dimensional *from/to* grid (switch *from* Treatment A *to* Treatment D) instead of a simple 1-dimensional allocation (X% of patients *reinitiating treatment* go to Treatment A, Y% to Treatment B, etc.) However, either of the Logit Choice algorithms can easily be extended to account for this complexity (2), and the

Table 9 Results of Applying ELC Formulation to Example Patient Flow of Reinitiating Treatment in the Standard Template

Total reinitiating treatment	2010	2011	2012	2013	2014	2015
Treatment A	2.3	1.8	2.3	3.3	3.3	5.0
Treatment C	16.1	12.9	4.4	3.3	3.3	5.0
Treatment D	0.0	0.0	0.0	3.3	12.0	5.0
Treatment BC	2.3	6.7	16.1	12.0	3.3	5.0
Treatment other	4.4	3.5	2.3	3.3	3.3	5.0
Total	50.0	50.0	50.0	50.0	50.0	50.0

Table 10 Example Treatment Change Matrix in Percentage Terms (2011)

From/to (2011)	A (%)	C (%)	D (%)	BC (%)	Other (%)	Total (%)
A	NA	40	0	30	30	100
C	30	NA	0	40	30	100
D	N/A	NA	NA	NA	NA	NA
BC	30	40	0	NA	30	100
Other	25	40	0	35	NA	100

NA, not applicable.

dynamic nature of Standard Template can then apply the resulting calculations to future time periods.

For example, assume that our hypothetical market research project also collected data on the dynamics of patients switching between treatment options in 2011. For simplicity, we will only look at one year of historical data, although such analyses often focus on more than one time period of switching dynamics. The example survey found the data shown in Table 10.

Table 10 details the distribution of patient switching dynamics as they move FROM individual treatment options TO other therapy regimens. For example, the data show that 40% of switches FROM Treatment A are TO Treatment C, 30% are TO Treatment BC, etc. Note it is not possible to truly "switch" from A to A, for example.* In this example case, we are investigating pure treatment switches. Also, note that TOTAL for all treatment switches must sum up to 100%—another advantage of having an MECE treatment option set.

Now we can use the Treatment Attractiveness scores for the MECE treatment option set to determine these cross option switching probabilities. The marketing literature is ripe with options for models dealing with such processes, often under the heading of first-order Markov processes. We have found the simplest and most accessible approach is to apply a Logit Choice algorithm (ALC or ELC) with a twist. For example, if a patient is to be switched from Treatment A, the distribution options are not a function of Treatment A's Treatment Attractiveness, but rather rely on the various utilities of the treatment options to which he or she might be switched. In effect, we can simply take Utility_A out of the equation and attempt to calibrate the switching probabilities with the remaining treatment attractiveness scores. To do so, we first make a relative Treatment Attractiveness matrix for 2011, using the previously presented utility data, as shown in Table 11.

Now the ALC can simply be applied for each potential TO element. For example, the ALC would predict that 36% of the patients switching FROM Treatment A would go TO Treatment C. (60/165 = 36%). In similar fashion, the entire grid can be filled out and compared to market-level data, as shown in Table 12.

* Titrations, as described in chapter 4, might be captured here in an easy extension of the model, however.

Table 11 Sample *From/to* Treatment Attractiveness Matrix (2011)

From/to	A	C	D	BC	Other	Total
A	NA	60	0	55	50	165
C	45	NA	0	55	50	150
D	45	60	NA	55	50	210
BC	45	60	0	NA	50	155
Other	45	60	0	55	NA	160

NA, not applicable.

Comparing the derived results shown in Table 12 with the sample treatment switching data in Table 10 shows that this algorithm calibrates to historical data quite nicely in this constructed example. We have found the modified treatment attractiveness matrix to be a useful framework for analyzing treatment changes in many of the markets we have investigated. However, there are cases in which the application of the ALC does not seem to fit collected data on switching distributions. In such situations, a number of options are available to the working team:

- Apply the ELC and compute a single x for the entire grid.
- Expand the ELC with a separate computed x for each treatment option being switched FROM $= x$ is unique to each row rather than the same for each treatment option.
- Try another choice model that better fits available data.

Regardless of the choice algorithm utilized, the iterative nature of dynamic models allows the relevant calculation to be repeated for each of several sequential time periods. In this example case, if we know from longitudinal analysis the absolute magnitude of patients switching FROM each of our defined treatment options, the dynamic model outputs would look something like Table 13 for 2011.

The values shown in Table 13 would change over the course of the dynamic model simulation due to differences in the absolute numbers of patients switching

Table 12 Sample *From/to* Treatment Change Matrix Based on Relative Treatment Attractiveness Scores in ALC Framework (2011)

From/to (2004)	A (%)	C (%)	D (%)	BC (%)	Other (%)	Total (%)
A	NA	36	0	33	30	100
C	30	NA	0	37	33	100
D	NA	NA	NA	NA	NA	NA
BC	29	39	0	NA	32	100
Other	28	38	0	34	NA	100

NA, not applicable.

Table 13 Results of Applying Modified ALC Formulation to Example Patient Flow of Switching Treatment in the Standard Template (2011)

From/to (2011)	A	C	D	BC	Other	Total
A	NA	9.1	0.0	8.3	7.6	25
C	10.5	NA	0.0	12.8	11.7	35
D	NA	NA	NA	NA	NA	N/A
BC	4.4	5.8	0.0	NA	4.8	15
Other	1.4	1.9	0.0	1.7	NA	5

NA, not applicable.

treatments and the application of the algorithm to changing Treatment Attractiveness values over time. In this manner, the flow of patients switching treatments is an evolving calculation based on the changing environment found in a given indication.

AN ALTERNATE APPROACH TO SWITCHING TREATMENT ALGORITHMS

In some pharmaceutical markets, evaluations of Treatment Attractiveness are either nonexistent, too difficult to estimate, or below the threshold of concern for the marketing team. Chapter 4 discussed how dynamic models can still be useful in such cases by allowing Patient Flow allocations to be input as scenario variables, instead of determined by integrating physician prescribing behavior and utility formulations in the full Standard Template. Another option is to utilize a first-order Markov model derive equilibrium market/patient shares. While basic Markov process formulations have certain drawbacks, on a number of occasions our clients have asked how the Dynamic Modeling process and associated projections compare to results derived from a first-order Markov model.

The mathematics behind Markov models can be somewhat intimidating, and will not be shown here to ensure continuity of the text.* In general, however, a first-order Markov model calculates equilibrium market/patient shares by extrapolating transition probabilities between individual products in the marketplace. In effect, the formulation focuses on product/treatment switching as the main driver of marketplace change over time, and provides an equation relating respective switching rates to long-run shares.

As such, Markov formulations often mirror the explicit operational structure of the *switching treatments* flow in a dynamic model. In fact, transition probabilities between various treatment options collected for a Markov model may be integrated directly into the stock/flow structure via a Treatment Change Matrix. This integration ignores the effect of Doctor Adoption and treatment attractiveness

* The interested reader can find full technical explanations of the Markov model formulation in Marketing Models by Lilien, Kotler, and Moorthy.

in the *switching treatments* algorithm, but does provide an easy way to leverage an established approach from the world of operations research. The results of the corresponding dynamic model can then be compared to the estimation of shares produced by a Markov model as a crosscheck for validity and robustness when evaluating an indication marketplace.

Integrating Utility, Choice Algorithms, and Constant Prescriber Status

The integration of Prescriber status, Treatment Attractiveness evaluations, and application of appropriate Logit Choice formulations produces a dynamic model which can now allocate three important flows in the Standard Template, as shown in Figure 3.

To this point in the progression, we have assumed that all physicians are Prescribers for each treatment option, now and in the foreseeable future. In the case of such constant and universal Prescriber status, each of the auxiliary variables allocating various patient flows are simply a function of the specific Logit Choice algorithms employed. Incorporating the entire Doctor Adoption dynamic

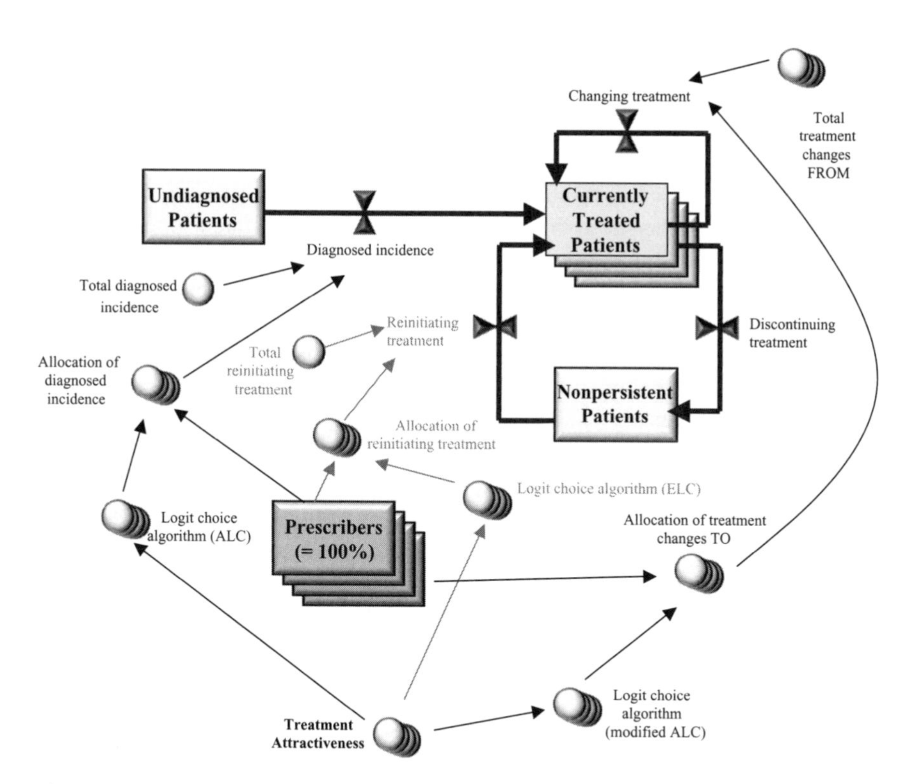

Figure 3 Inclusion of various Logit Choice algorithms with full Prescriber status in the Standard Template.

structure results in a more complete and realistic operational representation of key marketplace dynamics.

INCORPORATING DOCTOR ADOPTION/PRESCRIBING BEHAVIOR

We can now relax the assumption of constant and universal Prescriber status and incorporate a more dynamic picture of treatment adoption on the part of physicians. As chapter 4 demonstrated, the adoption of pharmaceutical treatment options by physicians has its own dynamics and structure that result in a changing number of physicians actively prescribing any particular option at any given time point. In aggregate, however, the key output from the stock/flow structure of Doctor Adoption is the number of Prescribers for each treatment option over time—usually expressed in terms of prescribing potential. In our continuing example, assume that the outputs from the Doctor Adoption structure look something like what is shown in Table 14.

A number of interesting points should be apparent from the set of sample outputs shown in Table 14:

- The percentage of physicians prescribing any of the MECE treatment option set is changing over time.
- For some treatment options, the prescribing percent is decreasing, mostly likely as doctors stop prescribing for one reason or another.
- The cases of Treatment A and Treatment D reflect launch dynamics in which the diffusion of doctors through the adoption process is apparent.
- Years in which the prescribing percent for a given treatment option is 0% reflects that *no* doctors are prescribing that particular therapy regimen, and hence *no* flows of patients will receive that treatment option in that time period. In some cases, this reflects the prelaunch period for a treatment option (D); and in others, it reflects doctors retrograding and disadopting a treatment option as a valid therapy option (BC).
- Prescribing propensity is shown in aggregate, but could be extended across dimensions of specialty and speed of adoption as per the description in chapter 8.

Table 14 Example Results of Doctor Adoption Dynamic Model Structure (2010–2015)

Percent prescribers	2010 (%)	2011 (%)	2012 (%)	2013 (%)	2014 (%)	2015 (%)
A	30	40	45	50	70	100
C	100	70	60	50	30	20
D	0	0	0	50	100	100
BC	70	60	55	50	10	0
Other	50	50	50	50	50	50

Table 15 Example Translation of Derived Share Results Using Prescriber Status (2013)

Treatment	Theoretical/model share (%)	Prescribers percent	Result (%)	Calculation	Normalized (%)
A	30	50	15	15/50	30
C	10	50	5	5/50	10
D	16	50	8	8/50	16
BC	40	50	20	20/50	40
Other	4	50	2	2/50	4
Total	100		50		100

With the concept and associated outputs from the Doctor Adoption structure in place, we can now integrate the three main sectors of the Standard Template in a combined approach within the Dynamic Modeling framework.

As we relax the assumption regarding universal doctor prescribing, the treatment attractiveness component (incorporating a Logit Choice algorithm) must somehow integrate with the percentage of doctors willing to prescribe any particular treatment option at any given time point. Recall that the Logit Choice returns the theoretical share if all doctors were active Prescribers of each treatment option. However, as we incorporate the Prescribers status outputs from Table 14, we must allow for the dynamics of adoption and variable prescribing behavior to enter the formulation. Marrying the two concepts is a two-step process, as follows:

1. Multiply each treatment option's theoretical share, as determined by the ALC or the ELC, by each treatment option's status in terms of prescribing behavior.
2. Normalize the resulting calculations to ensure a total of 100%.

This translation process simply accounts for Prescriber status and applies it to the theoretical shares as determined from a Logit Choice algorithm. For example, assume that a simple ALC formulation (under conditions of 100% Doctor Adoption status) produced a set of Theoretical/Model share results for 2013, which are then modified by the Percent Prescriber results from Table 14. The example translation methodology is shown in Table 15.

Note that the total of the Result column in Table 15 does *not* sum up to 100%—a troublesome result in a modeling effort designed to account for and appropriately direct individual Patient Flows. But by normalizing the Result column, the translation process depicted in Table 15 provides the ability to fully allocate the any aggregate flow of patients across the elements of the MECE treatment option set.

Note that if the Prescribers percent is identical for each treatment option (in the case depicted in Table 15, 50% of the physicians are prescribing each of the five MECE treatment options in 2013), when Normalized, the Result reduces to exactly the Theoretical/Model Share outputs. This is true for any Prescriber

Table 16 Example Results of Combining Derived Share Calculations with Prescriber Status (2010–2015)

Treatment	2010 (%)	2011 (%)	2012 (%)	2013 (%)	2014 (%)	2015 (%)
A	18	27	31	30	46	60
C	20	16	14	10	7	4
D	0	0	0	16	35	32
BC	57	53	51	40	9	0
Other	4	4	5	4	4	4
Total	100	100	100	100	100	100

percent status shared by all treatment options—up to and including 100%, which is justified given our assumptions about Treatment Attractiveness and the outputs of the Logit Choice algorithm.

For simplicity, assume that the Theoretical/Model Share depicted in Table 15 never changes over time—i.e., each treatment option's utility is constant and therefore the computed share for each therapy regimen is unvarying as well. Applying the same translation process of combining the Doctor Adoption results from Table 14 for the entire 2010–2015 timeframe in this continuing example results in flow allocation numbers as shown in Table 16.

Because the Logit Choice algorithm is returning constant outputs for each treatment option, the variability in shares shown in Table 16 is due solely to the changes in Prescribers status as determined by the Doctor Adoption structure. Of course, in the full Standard Template, Theoretical/Model Share outputs are changing over time due to variations in treatment attractiveness scores and entry of pipeline products, providing a much more realistic representation of evolving marketplace dynamics.

BRINGING THE THREE SECTORS TOGETHER

By integrating the three sectors of Patient Flow, Doctor Adoption, and Treatment Attractiveness, the Standard Template can now account for combined dynamics producing behavior in a given indication marketplace over time. In addition, the methodology operationally identifies key strategic levers which might inform or direct the marketing approach of a given product to an indication, as shown in Figure 4.

Chapter 12 will demonstrate how a fully integrated Standard Template model, complete with data parameters and calibrated to historical data as has been described to this point, can be used in the Brand Planning process for scenario testing and strategy development. Many of our clients use such dynamic models on an ongoing basis as they evolve their strategic thinking regarding indication marketplaces and the effective ways to operate in them.

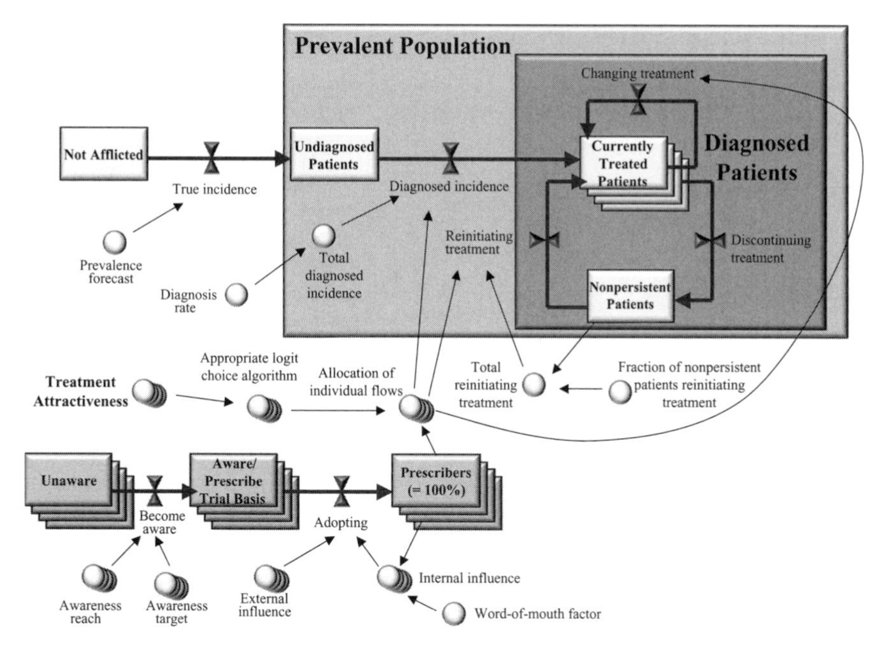

Figure 4 Fully integrated Standard Template dynamic model.

SUMMARY

- Logit Choice algorithms are mathematical formulations to translate the utility scores for various treatment options into derived share numbers for those therapy regimens.
- Different Logit Choice formulations can be applied to different parts of the Patient Flow structure and calibrated to specific historical data.
- Introducing the Doctor Adoption sector results in a fully integrated Standard Template dynamic model that operationally defined key interrelationships driving the behavior of pharmaceutical marketplaces.

REFERENCES

1. Lilien G, Kotler P, Moorthy KS. Marketing Models. Englewood Cliffs, NJ: Prentice-Hill, 1992.
2. Rust R, Lemon K, Zeithaml V. Driving Customer Equity: Linking Customer Lifetime Value to Strategic Marketing Decisions. Marketing Science Institute Working Paper 01–108, 2001.

12

Strategy Testing Using Integrated Standard Template Model

INTRODUCTION

This chapter will cover the following topics:

- Organizing framework for New Chemical Entity (NCE) introduction and expected sales trajectory
- The role of Patient Flow, Doctor Adoption, and Treatment Attractiveness in the organizing framework
- Simulation tests using the Standard Template dynamic model
- Strategic implications of simulation scenario tests

STRATEGY FRAMEWORK FOR NEW COMPOUND INTRODUCTIONS

Pharmaceutical markets are complex and analyzing them in the context of strategic development is often a difficult proposition. A number of organizing principles are common throughout the pharmaceutical world—Porter's Five Forces and the ubiquitous Strengths, Weaknesses, Opportunities, and Threats (SWOT) analysis are widely accepted approaches for framing strategic issues. In our consulting practice we have developed an operating framework that provides a useful way to think about marketing strategy when entering/analyzing an indication marketplace, as shown in Figure 1.

1. The **Case** for a new compound is the set of aggregate, macroeconomic market conditions of the indication being evaluated. Information on epidemiology, changing demographics, treatment paradigms, historical market revenue trends, and existing competitors all define the Case for an NCE considering entry into a new marketplace. Case issues are often summarized as revenue forecasts for an indication (see chap. 1), although by now readers should readily recognize the underlying operational dynamic assumptions behind such

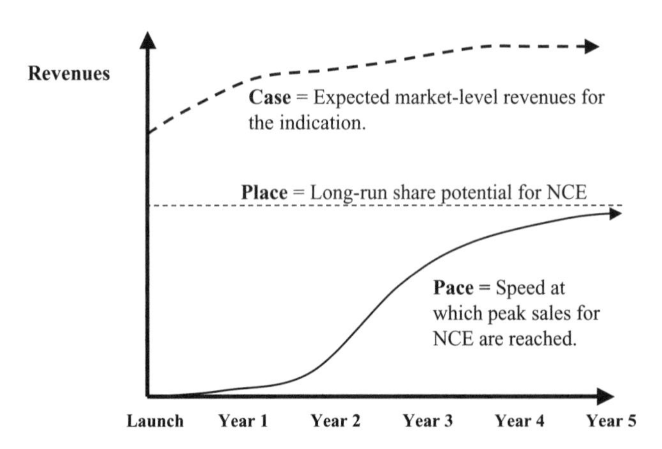

Figure 1 Case, Place, and Pace framework an NCE in a given indication.

projections. The concept of Case is vital when evaluating the feasibility of continued support of a developmental compound, as it determines the aggregate conditions into which such a product introduction will take place. The strategic influence of individual pharmaceutical firms on case issues must be carefully evaluated, however, as by definition strategies designed to change the Case redefine the market environment for all players in it—resulting in a larger pie for collaborators and competitors alike.

2. The **Place** for a new compound defines how it compares to the existing and future competition in terms of efficacy, safety, side effects, etc. In markets exhibiting specific treatment algorithms, Place may also include where the compound is expected to fall within a defined Lines of Therapy progression. The place for an NCE is in some ways determined by its U.S. Food and Drugs Administration-approved label which can limit a compound's prescribed usage to certain segments of the patient population, based on metrics such as disease severity or failure of existing treatment options to control the disease. In economic lingo, the Place of a compound in an indication determines its long-run market share potential—often referred to as peak sales in the pharmaceutical world. This upper bound of commercial potential for a compound is under the influence of marketing efforts, however, as increasing a compound's perceived Treatment Attractiveness or pursuing a labeling strategy to get the compound earlier in a Line of Therapy progression can impact where that long-run share ceiling is observed. The specific design of clinical trials may also factor into the equation, as the outcomes of such trials are often key determinants in labeling and perception of NCE utility in the existing marketplace.
3. The **Pace** for the new compound relates to the speed at which it can reach peak share potential—the shape of the expected sales trajectory (see chap. 1). While Case and Place are generally functions of product attributes, disease epidemiology, and other aggregate conditions, the Pace for an NCE is more commonly

under the influence of various strategic marketing decisions of the part of individual pharmaceutical firms. Effective Brand Planning leverages the observed dynamics of Patient Flow and Doctor Adoption to effectively increase the Pace at which an NCE is accepted in the marketplace. The resulting expected sales trajectory, commonly referred to as product uptake, is an important factor in the commercial assessment of the compound using financial metrics such as Net Present Value (NPV).

RECAP OF EXAMPLE DYNAMIC MODEL MARKET CONDITIONS

Throughout this text we have presented a detailed extended example of how an integrated Standard Template dynamic model can be created by combining key aspects of patient dynamics, physician acceptance of new prescription medications, and the methodologies by which those treatment options can be evaluated. This chapter will introduce how such an integrated model can be used for Brand Planning and strategic development as it relates to the Case, Place, and Pace for a particular NCE.

Recall our extended example focuses on a chronic indication with a limited number of existing therapy options and a new compound slated to enter the marketplace at the beginning of 2013. The key Patient Flow parameters at an aggregate level are as shown in Table 1 (see chaps. 4 and 10).

Chapter 8 described how a detailed Dynamic Modeling analysis of the launches of previously released treatments in an example indication allowed for the strength of various marketing levers on the adoption of NCEs by prescribing physicians can be developed. For Treatment D in our continuing example, assume that such a Doctor Adoption model was developed for this indication, and the results of expected awareness targets and marketing spending to support Treatment D projected its Prescriber status as indicated in Table 2.

In addition, the scoring exercise needed to determine Treatment Attractiveness was described in chapter 10, and chapter 11 detailed how the results of the determination of the relative utilities of various therapy options in the indication marketplace could be applied. Of particular interest is the utility evaluation for Treatment D, as shown in Table 3.

Table 1 Aggregate Patient Flow Parameters in Example Standard Template Dynamic Model

Patient flow	2010	2011	2012	2013	2014	2015	2016	2017
True incidence	23	36	29	26	26	38	39	39
Diagnosed incidence	16	29	22	19	19	28	29	29
Switching treatments	80	82	89	95	109	133	158	178
Discontinuing treatment	20	23	25	27	29	32	35	37
Reinitiating treatment	25	24	23	24	24	25	27	28

Table 2 Expected Prescriber Status from Doctor Adoption Sector of Standard Template Dynamic Model

End of calendar year:	2012 (%)	2013 (%)	2014 (%)	2015 (%)	2016 (%)	2017 (%)
Prescriber status: treatment D	0	40	85	95	97	97

Models are not built for the sake of modeling itself, however, and the pertinent questions center on how to use such an integrated dynamic model to build a consensus understanding of an indication marketplace and how simulation analysis can help to develop more effective strategies therein. There are as many ways to test models as there are to build them, but our clients have found the Case, Place, and Pace framework to be a useful a guiding principle.

First, a **Most Likely Scenario** for the dynamic model simulation results needs to be established. A distribution of possible outcomes is generally produced through simulation with the Most Likely Scenario representing the most commonly observed results, but for the sake of simplicity we will show the Most Likely Scenario as a set of single set of model outputs related to the number of patients on Treatment D at the end of each of three years postlaunch, as shown in Figure 2.

Using the Dynamic Model to Analyze the Case for a New Compound Introduction

Operational models such as the continuing example throughout this text can be used to analyze the aggregate conditions into which an NCE might be launched. Such evaluations of the Case for a new compound provide a way to test the feasibility of market entry, and also assess the impact of various environmental situations with might be under the influence of a given pharmaceutical firm.

Various types of model analysis using Dynamic Modeling simulations are described throughout this text, including varieties of sensitivity analysis (chap. 2), Monte Carlo modeling (see chaps. 14 and 16), and optimization under uncertainty (see chap. 14.) For the purposes of continuity, the analysis in this chapter will show the results of basic parameter sensitivity analysis, although other techniques can certainly be applied as well.

Table 4 shows the results of various scenario tests run through the example dynamic model to observe their impact on the trajectory of patients on Treatment D.

Table 3 Derived Treatment Attractiveness for Treatment D

	2012	2013	2014	2015	2016	2017
Treatment attractiveness: treatment D	0	50	60	50	50	50

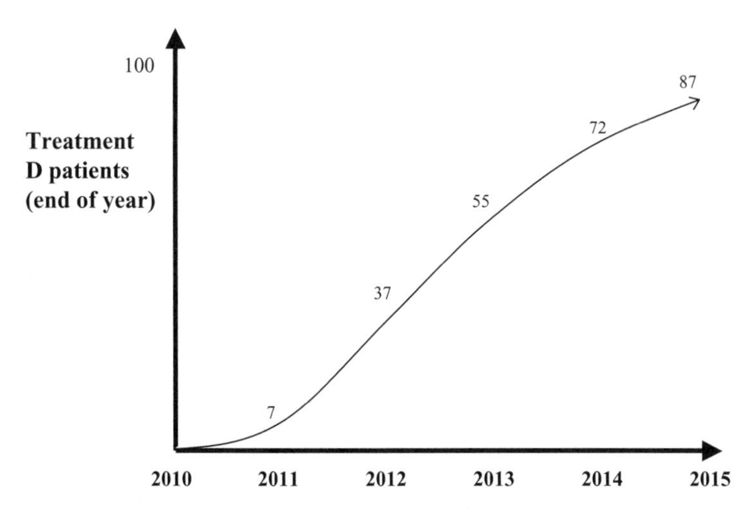

Figure 2 Most Likely Scenario of Treatment D Patients from example integrated Standard Template dynamic model.

Scenarios such as the ones shown in Table 4 begin to give some clues as to how the Case for a new compound is affected by macroeconomic and disease conditions. For example, epidemiology for a disease is never fully accurate, so sensitivity tests on metrics such as prevalence help inform a range of possible market conditions. Similarly, pharmaceutical drugs are sometimes pulled from the market due to adverse side effects, and a scenario in which Treatment BC exits the marketplace in 2013 can be modeled fairly easily. Other scenarios determining the Case for an NCE are more likely to be influenced of the firm launching Treatment D. Accelerating the expected launch time by three months, all else being equal, would translate into a more rapid uptake but a similar number of patients on the treatment at the end of 2015 when compared to the Most Likely Scenario. Another interesting result occurs when the importance of Efficacy is 20% higher in the minds of practicing physicians in determining relative Treatment Attractiveness, resulting in about a 10% increase in the number of patients on Treatment D at each year.

Table 4 Example Dynamic Modeling Simulation Results Related to the Case for Treatment D

Scenario—treatment D patients	2012	2013	2014	2015	2016	2017
Most likely scenario	0	6.6	36.9	55.2	72.4	86.9
Prevalence up by 5%	0	6.9	38.7	58	76	91.2
Launch time 3 months earlier	0.3	9.5	40.7	57.9	74	87.9
Treatment BC exits market 2013	0	6.6	36.9	76.4	91.7	103.1
Efficacy 20% more important	0	7.1	40.3	60.6	79.3	95

Strategic Implications of Sensitivity Analysis Related to the Case for an NCE

The results of scenarios regarding the Case for an NCE give insight into possible market conditions, but in some cases suggest strategies designed to impact the overall indication marketplace. In the case of epidemiology, new standards of inclusion for disease categorization could be pursued, such as lowering the blood pressure thresholds associated with a diagnosis of hypertension or expanding the range of behaviors classed as having a certain psychiatric disorder. Launch times could be influenced by more efficient development processes that are larger or concurrent phase trials or better filing procedures to eliminate rework. The financial feasibility of such actions can be determined by comparing associated costs to the expected corresponding revenue gains (see chap. 13). And although the withdrawal of a competitor is clearly not within the sphere of influence, such an event could potentially change the commercial viability of a compound. Furthermore, the importance of efficacy in the prescribing decision could be influenced by marketing campaigns to physicians emphasizing the need for control of disease conditions.

In these ways, the Dynamic Modeling evaluation of the Case for an NCE provides an operational tool to not only assess conditions of the potential indication marketplace but also to analyze the impact of changing various macro inputs on the Case situation. The strategies resulting from such analyses are likely to be more robust than those based on untestable mental models of the Case for a new compound.

Using the Dynamic Model to Analyze the *Place* for a New Compound Introduction

To help determine the Place for a given NCE, the sensitivity of Dynamic Modeling results based on various levels of calculated utility for that therapy option can be analyzed. These results can be assessed in aggregate using the overall Treatment Attractiveness for the new compound or by the individual components of computed utility, as shown in Table 5.

Table 5 Example Dynamic Modeling Simulation Results Related to the Place for Treatment D

Percent increase in treatment D patients above the Most likely scenario	2015 (%)
Treatment attractiveness (treatment D) up by 10%	14
Safety rating (treatment D) up by 10%	1
Efficacy rating (treatment D) up by 10%	9
Onset of action rating (treatment D) up by 10%	3
Side effects rating (treatment D) up by 10%	2

Strategic Implications of Sensitivity Analysis Related to the Place for an NCE

Sensitivity analysis results such as those shown in Table 5 should suggest areas where strategic marketing initiatives and/or design of clinical trials could be used to position a new compound effectively. Perhaps Treatment D suffers from a perceived quality deficiency in which the perception of the drug by prescribing physicians lags behind empirical or clinical evidence. A strategy of physician education might be enacted to correct this perception delay. And the results shown in Table 5 clearly show that the most important component of Treatment Attractiveness is the rating of Treatment D in terms of efficacy. A phase III (prelaunch) or phase IV (postlaunch) clinical trial focusing on collecting data related to the effectiveness of Treatment D might increase the rating of that drug in the eyes of doctors treating patients in the indication.

The other key component of the Place for an NCE involves its position in a defined line of Therapy progression. Modeling this dynamic was explaining in chapter 5 but was not included in our continuing example of the Standard Template. Including Lines of Therapy structure into the Patient Flow sector is a relatively easy extension, and allows for this dynamic to be a strategic possibility when evaluating the Place for a new compound.

Using the Dynamic Model to Analyze the *Pace* for a New Compound Introduction

The Pace for a given NCE is driven by dynamics, and its evaluation can be made by comparing the relative magnitudes of Patient Flow and Doctor Adoption in determining the expected sales trajectory for the compound. This analysis looks at the Most Likely Scenario versus various levels of Patient Flow and Doctor Adoption Pace to determine their impact on the product uptake of Treatment D, as shown in Figure 3.

Figure 3 shows that increasing the magnitude of the Patient Flows in the example indication is more important than changing the rate at which physicians are expected to adopt the NCE, all else being equal. The increased Patient Flow scenario assumes that the flows of diagnosed incidence, switching treatment, and reinitiating treatment are 10% greater than in the Most Likely Scenario. The increased Doctor Adoption scenario simply takes the derived diffusion metric shown in Table 2 and bumps it up by 10%—subject to a maximum of adoption of 100%. The analysis shows that increasing the rate of Doctor Adoption certainly buys Treatment D a few additional patients, but fewer than result from an increase of the Patient Flow by the same amount.*

* This result is not true in all cases, and in part results from the rather aggressive assumption of Doctor Adoption in the Most Likely Scenario. We have seen many real-world indications in which increases in physician acceptance were the key determinant in the successful launch of a new compound.

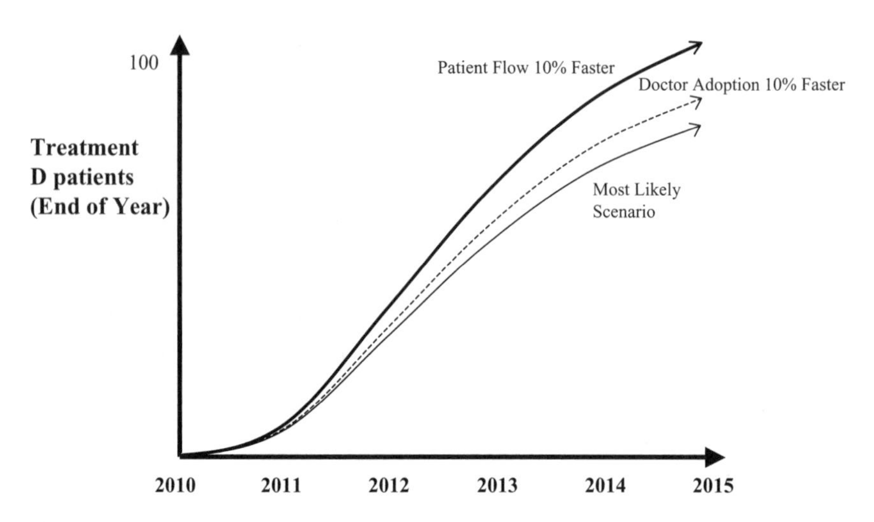

Figure 3 Impact of increased Patient Flow and Doctor Adoption on the Most Likely Scenario of Treatment D patients.

Strategic Implications of Broad Patient Flow vs. Doctor Adoption Scenarios

Results such as those shown in Figure 3 begin to focus the strategic development effort on the broad dynamic categories affecting the Pace for an NCE. This example case indicates that at an aggregate level, increases in the Patient Flow dynamics have a larger impact on product uptake than corresponding increases in Doctor Adoption above the Most Likely Scenario. These effects can then be broken down into the more detailed components driving the Patient Flow and Doctor Adoption sectors from the Standard Template.

Patient Flow Metrics

The detailed components of the Patient Flow sector of the Standard Template can also be analyzed to determine their relative impact on the Pace of new compound uptake. As has been demonstrated throughout this work, the number of patients on any given treatment can be converted into revenues using metrics such as true compliance and an average price per patient day of therapy (see chap. 4). The resulting revenue projections are useful from a financial standpoint as another methodology to evaluate the commercial potential of an NCE, but also provide a practical means to analyze model sensitivity results. Revenue streams can be discounted back to the current timeframe using an appropriate discount rate to arrive at a ***Cumulative Discounted Revenue*** (CDR) calculation, which provides a useful metric when comparing the results of various Dynamic Modeling simulations.

In our continuing example, suppose such an CDR calculation was performed on the Most Likely Scenario for Treatment D, with the resulting CDR being

Table 6 Example Dynamic Modeling Simulation Results Related to the Impact of Patient Flow Parameters on the Most Likely Scenario for Treatment D

Scenario	CDR—treatment D ($ M)
Most likely scenario	165
Diagnosed incidence flow 10% greater	181
Switching from existing drugs 10% faster	175
Switching from treatment D 10% slower	169
Time as nonpersistent 10% less	168
Discontinuation from treatment D 10% slower	166

$165M. The sensitivity of this metric on various changes in aggregate Patient Flow metrics can then be compared, as shown in Table 6.

The sensitivity analysis results shown in Table 6 indicate that the flow of *diagnosed incidence* has the greatest impact on the ultimate commercial potential of Treatment D, with the parameters controlling the dynamics of switching from existing treatments falling running a close second. The Patient Flow dynamics associated with switching from Treatment D, treatment discontinuation from all forms of therapy, and reinitiation of treatment after being Non-Persistent appear to have less impact on the overall behavior of the system.

Strategic Implications of Patient Flow Sensitivity Analysis

The simulation results such as those seen in Table 6 give strategic insight into leverage points within the Patient Flow sector of a dynamic model for a particular disease indication. First, the magnitude of the sensitivity analysis (10% in this example case) should be evaluated, as a 10% change in switching rates may be easier bring about than a 10% change in the flow of newly diagnosed patients. Second, the expected marketing costs associated with enacting these parameter changes can be estimated and compared with the resulting CDR metric to determine financial feasibility of strategic options. The Strategic Sensitivity Grid (see chaps. 2 and 7) provides a useful framework for analyzing these tradeoffs. For example, increasing the diagnosed incidence flow results in an additional $16M in CDRs for Treatment D ($181M–$165M in the Most Likely Scenario) but such an action is not fiscally viable if the cost to do so is estimated at $40M. In this manner, Patient Flow Sensitivity Analysis can identify potential points of strategic leverage and also evaluate the financial implication of marketing strategies when using the CDR metric.

Patient Flow Allocations

In addition to analyzing changes in the magnitude of Patient Flows, the dynamic model can be used to test scenarios involving the allocation of those flows across

Table 7 Example Dynamic Modeling Simulation Results Related to the Impact of Patient Flow Allocation on the Most Likely Scenario for Treatment D

Percent increase in most likely scenario of treatment D patients	2015 (%)
Treatment D's share of newly diagnosed patients up 10%	2
Treatment D's share of switching patients up 10%	2
Treatment D's share of reinitiating patients up 10%	9

various treatment options—particularly Treatment D. Recall from chapter 10 that the elements of treatment attractiveness and Doctor Adoption can be combined in a Logit Choice algorithm to make these allocation calculations in the integrated Standard Template. Simulations tests can be run that override this algorithm, however, to test the impact of Treatment D garnering more than its calculated fair share of various flows of patients, as shown in Table 7.

In this example indication, the flow of patients reinitiating treatment is not strategically important but increasing the share of those patients captured by Treatment D has the greatest impact on the final number of patients. Table 7 shows the results of various Patient Flow allocation scenarios at a 10% sensitivity level, but the magnitude of these scenarios could be tested over a wide range of possibilities to test the impact of scale of effect on overall simulation results.

Strategic Implications of Patient Flow Allocation Sensitivity Analysis

Simulation results such as those seen in Table 7 identify strategic leverage points related to the places where capturing patients is most vital to commercial success. In this example indication, these results suggest that getting a patient on Treatment D first (share of newly diagnosed patients) or as they switch therapies (share of switching patients) is not as important as capturing Non-Persistent patients as they return to treatment (share of reinitiating patients. A strategy developed to tap into this flow of patients reinitiating treatment is therefore more likely to have a positive impact on the overall performance of Treatment D in the indication marketplace.

Combining Patient Flow Metrics and Allocation

Dynamic model simulation results are especially interesting when combinations of strategies are tested and evaluated, and these results sometimes show synergies which would not be expected from analyzing individual strategies in isolation. Recall that the combination of dynamics in the Patient Flow sector is analogous to a stream of water from a drinking fountain. Aggregate metrics determine how big the flows of patients are, and allocation algorithms determine how much of a "sip" each treatment option is able to take. Combinations of strategies involving aggregate flows and allocation of patients can be thought of as increasing the flow of water coming from the drinking fountain *and* letting a particular therapy take a bigger sip, as shown in Table 8.

Table 8 Comparison of Algebraic and Simulation Results for a Combination of Strategies Involving Patient Flow and Allocation on the Most Likely Scenario for Treatment D

Incremental treatment D patients above the most likely scenario	2012	2013	2014	2015	2016	2017
Increase switching from existing drugs by 10%	0	0.4	2.1	3.4	4.5	5.3
Increase treatment D's share of switches by 10%	0	0.8	3.3	5.2	6.9	8.3
Algebraic results of combination of strategies	0	1.2	5.4	8.6	11.4	13.6
Simulation results of combination of strategies	0	1.4	5.8	8.9	11.8	14.1

Table 8 indicates that simply adding up individual strategies does not produce the same results as inputting those strategies into the integrated dynamic model and simulating the results. Simulations are iterative; with each parameter change, dictating a new set of conditions from which the results for the next subsequent time period can be calculated. As such, the scenarios involving increased switching from existing drugs and an increase in Treatment D's share of switches are complimentary strategies, resulting in more Treatment D patients through a combination of dynamic factors around the switching dynamics. Combination strategies such as depicted in Table 8 *must* be evaluated using simulation, as dynamic models take into account all of the relevant impacts of such strategies in an integrated framework.

Strategic Implications of Patient Flow Combination Strategies

Simulations evaluating the effects of combinations of strategies should not be performed in an ad hoc fashion, and the operational nature of the Standard Template should suggest some inherently synergistic strategies to be tested. For example, a marketing program to more likely to encourage patients or doctors to change to a particular therapy (Treatment D) than simply switch to other therapy for the sake of just switching. Similar combinations of Patient Flow strategies might address Non-Persistent patients ("return to therapy and try Treatment D") or newly diagnosed patients ("if you have these symptoms, ask your doctor about Treatment D"). The effects of these types of clear combination strategies can then be rigorous evaluated through Dynamic Modeling simulations.

Doctor Adoption Metrics

Similar to assessing the impact of the more detailed components of Patient Flow, component factors of the Doctor Adoption sector of the Standard Template can be analyzed to determine the relative impact on the Pace of new compound uptake.

Table 9 Example Dynamic Modeling Simulation Results Related to the Impact of Doctor Adoption Parameters on the Most Likely Scenario for Treatment D

Number of treatment D patients	2012	2013	2014	2015	2016	2017
Most likely scenario	0	6.6	36.9	55.2	72.4	86.9
Physician awareness 10% higher	0	7.4	38.4	56.3	73.1	87.3
Physician marketing 10% higher	0	7	38	56	72.9	87.2
Physician WOM 10% higher	0	6.8	38	56.3	73.2	87.4

Chapter 8 detailed the Dynamic Modeling approach to analyzing the acceptance of newly released pharmaceutical treatments, including elements of physician marketing. For the sake of simplicity, Table 9 shows the results of a sensitivity analysis on three aggregate metrics of doctor adoption such as awareness, physician word of mouth (WOM), and aggregate marketing spending.

The results of Table 9 show that in this example case, the impact of various parameters affecting Doctor Adoption and subsequently the numbers of patients on Treatment D is remarkably similar. This outcome suggests that the relative strength of external factors such as awareness and physician marketing and internal factor such as WOM are almost identical in their effect on the expected sales trajectory for Treatment D in our continuing example.

Strategic Implications of Doctor Adoption Sensitivity Analysis

Simulations evaluating the effects of various parameters affecting the acceptance of new compounds by physicians can focus strategic effort on places of high leverage. In this example case, the relative impact of various marketing levers appears to be almost identical; so, the allocation of resources between various types of physician marketing campaigns is not entirely clear. In other market situations, such an analysis might identify key areas of strategic influence where Doctor Adoption parameter changes have a more pronounced impact on compound uptake. Resources could then be directed to those areas to maximize the expected sales trajectory of a given NCE.

Sensitivity analysis can also be used to determine the effect on a model metric based on various levels of a given input. For example, WOM parameters are notoriously difficult to estimate so a range of input values are often tested to assess the magnitude of their impact, as shown in Figure 4.

Figure 4 clearly illustrates that the existence of WOM plays a big role in the expected Pace for a new compound, as the simulation results show that the expected sales trajectory for Treatment D is significantly lower in the case where WOM is nonexistent. Increasing the WOM parameter above Most Likely Scenario value does have a positive impact, the effects of which can be analyzed by examining the corresponding product uptakes more closely. Analyzing the magnitude sensitivity of individual parameters such as WOM across a wide range of values can help identify the importance of reasonable estimates on overall simulation results as described in chapter 13 in detail.

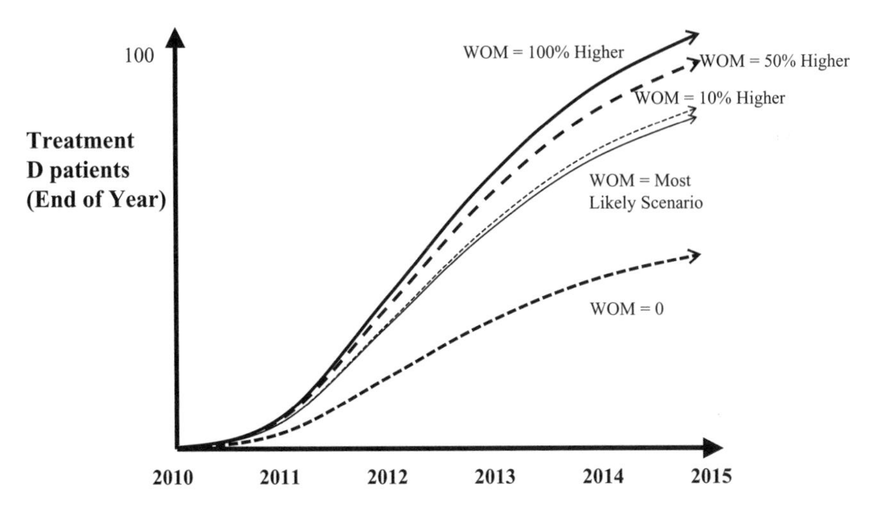

Figure 4 Impact of various magnitudes of WOM on the Most Likely Scenario of Treatment D patients.

COMBINING CASE, PLACE, AND PACE STRATEGIES

As has been discussed throughout this text, the Dynamic Modeling framework provides an operational way to test the impact of various marketing strategies on the expected performance of an NCE in an indication marketplace. Individual parameter tests such as those described in this chapter are useful in determining overall strategic direction of Brand Plan marketing campaigns. But individual strategies can be combined into an integrated strategic plan that addresses issues of Case, Place, and Pace in an indication market.

For example, a marketing team for Treatment D might want to know the combined effects of launching the compound three months earlier (Case), having the Treatment Attractiveness for D be 10% higher (Place), and increasing the speed at which physicians will adopt the new NCE (Pace). These scenarios can be run in isolation using the dynamic model simulation, but the combination of effects cannot be determined using a mental model or by simply adding up the impacts of the individual strategies by themselves. Combinations of strategies can be simulated very easily, however, and the results of this example are shown in Figure 5.

Strategic Implications of Combinations of Strategies Involving Case, Place, and Pace

Simulations evaluating combinations of strategic initiatives prove very useful when analyzing the possible sales trajectories for a new compound. These strategies are usually derived from careful analysis of individual sensitivity results such as those described in this chapter. But the integrated Dynamic Modeling methodology allows these strategies to be merged into a coherent strategic plan that encompasses

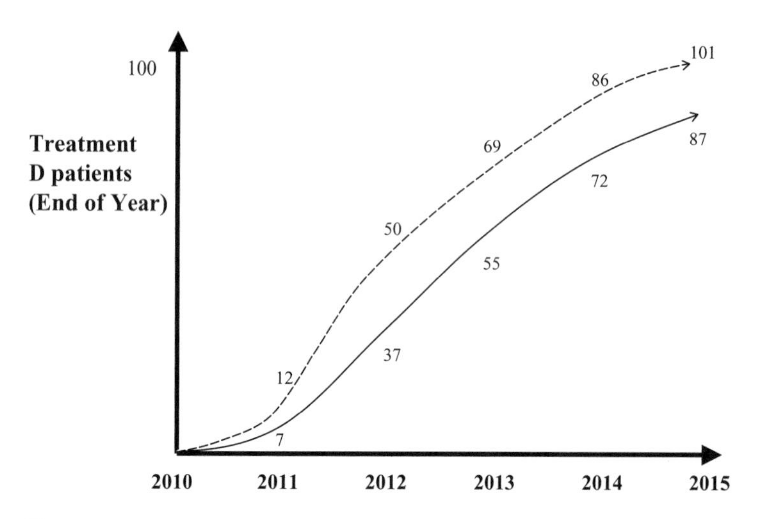

Figure 5 Impact of combined strategy of Case, Place, and Pace on the Most Likely Scenario of Treatment D patients.

the overall conditions of the indication market, the position of the NCE within it, and the speed at which the compound will be accepted into the marketplace. The results of this analysis allow a Brand Planning team to evaluate the feasibility of the expected sales trajectory for an NCE determined from traditional forecasting methodologies, as well as develop an integrated strategy for compound positioning and its supporting marketing plan. This process has allowed many of our clients to more effectively allocates resources related to the Case, Place, and Pace for a new compound introduction.

SUMMARY

- The Case, Place, and Pace framework is a useful approach to thinking about the market environment, the position of an NCE within it, and the shape of the expected sales trajectory of the compound.
- The Patient Flow, Doctor Adoption, and Treatment Attractiveness sectors of the Standard Template are designed to support the framework by operationalizing the concepts of Case, Place, and Pace.
- Dynamic model simulations generate a most likely scenario of results, from which a series of scenarios can be run to develop effective strategies for a Brand Planning.
- Sensitivity analysis can determine the absolute impact of single parameters on the marketplace, the magnitude of various levels of inputs, and the effects of combinations of strategies.
- Effective Brand Plans can be informed by careful testing and understanding of Dynamic Modeling results, particularly in the context of the Case, Place, and Pace for a developmental compound.

13

Dynamic Modeling of First-Mover Advantage and Network Effects

INTRODUCTION

This chapter will cover the following topics:

- Financial considerations associated with accelerated New Chemical Entity (NCE) development
- Reinforcing feedback mechanisms in pharmaceutical marketplaces
- Introducing network effects into the Standard Template
- Three periods of first-mover advantage
- Using Dynamic Modeling to account for network effects
- Using simulation to quantify first-mover advantage

ACCELERATED NCE DEVELOPMENT

Senior managers in pharmaceutical firms often face a difficult trade-off between cost and speed in new product development. Allocating additional resources to developing a compound for a specific therapeutic area can accelerate the process, but usually at a significant cost. For example, compound development times can be reduced by recruiting additional patients for clinical trials, running multiple trials to reduce the risk of the U.S. Food and Drug Administration rejecting subsequent results, or developing multiple compounds in parallel to address a given indication with the expectation that only one or two will be successful. In fact, some researchers argue that pharmaceutical companies often start too few new product approaches in a therapeutic category, given high rates of compound failure in the development process (1).

Conventional industry wisdom says that shorter development times provide the advantages of being first to market with a new product—a view supported by recent research (2,3). Even in indications crowded with existing products, pharmaceutical firms often race to beat the competition to the marketplace. But shorter

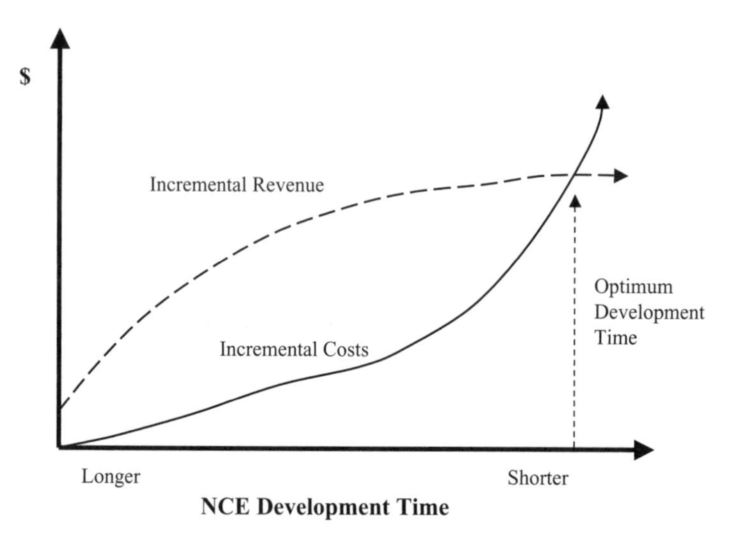

Figure 1 Incremental costs and revenue tradeoffs in NCE Development times.

development times generally require higher development costs—a balancing act that is especially difficult if development spending is constrained by capital availability, or if faster development of one compound reduces the resources available for other projects. Analyzing the trade-offs between development costs and time to market is essential for effective resource allocation and strategy development, and the Dynamic Modeling framework provides a rigorous methodology for analyzing such situations.

In order to effectively quantify the pros and cons of accelerated NCE development; an operational evaluation of the indication marketplace dynamics must be developed. As previous chapters have demonstrated, dynamic models create a simulation environment in which these types of "what if" questions can be addressed. The simulated marketplace provides a way to test various scenarios, including the costs and potential benefits of being first to market in an indication.

The optimal pace of product development can be determined by considering the incremental costs and benefits of greater speed to market. Figure 1 plots the marginal revenue and associated costs of accelerated NCE development.*

Figure 1 visually represents the trade-offs facing pharmaceutical firms as they balance NCE development speed versus associated costs. Shorter development times are more costly but correspond to additional revenues for the compound, for at least two key reasons.

First, faster development directly increases the time the NCE is both on the market and under patent protection. The clock starts ticking for the NCE

* Readers with nostalgia for their college microeconomics class will recall the equilibrium point in Figure 1 simply represents the point at which marginal costs equal marginal revenues—the solution which maximizes profit.

from the point of patent filing, and the more time spent on the market generating revenue during the 20-year patent protection period translates to additional cumulative revenue for the compound and the brand. Revenue falls significantly when compounds come off-patent; witness the wave of mergers, acquisitions, and compound in-licensing on the part of major pharmaceuticals in the first part of the 21st century as they face patent expiry of a number of key compounds.

Second, reinforcing feedback processes or "network effects" create the possibility of significant and sustained advantages to being the first to market. Chapter 3 introduced the concept of feedback—a set of interrelationships in which the status of a system condition enacts a circular chain of driving forces which eventually "feedback" to affect the original variable. Feedback relationships sometimes take the form of reinforcing relationships such as the interest-bearing checking account in chapter 3. The greater the account balance, the higher the interest earned, which pushes the account balance even higher. These types of relationships form a reinforcing feedback scenario in which gains produce greater gains through a circular set of interrelationships.*

REINFORCING FEEDBACK EFFECTS

Reinforcing feedback effects are sometimes seen in pharmaceutical markets and represent a "virtuous cycle" in which marketing efforts and word-of-mouth push product acceptance and diffusion beyond conventional expectations. The explosion in sales and brand recognition of Viagra is the best-in-class example of product "take off" that cannot be described in simple linear terms.† Such stratospheric performance generally eclipses Expected Sales Trajectories (see chap. 1) and are the product of reinforcing feedback mechanisms that may be at play in any particular indication marketplace.

Common Reinforcing Feedback Mechanisms in Disease Marketplaces‡

Three common reinforcing feedback mechanisms in the pharmaceutical industry are as follows:

1. *Direct physician communication*: Positive experiences with a particular drug, often encourages physicians to inform their colleagues about its benefits. The process is not dissimilar from the word-of-mouth processes that drive unexpected successes in other realms such as "sleeper hits" from the movie industry like *My Big Fat Greek Wedding* or *Juno*. Strong word-of-mouth

* Reinforcing feedback relationships in economics and the business world have been popularized by Dr. Brian Arthur and others at the Santa Fe Institute under the heading of Increasing Returns. See Increasing Returns and Path Dependence in the Economy.

† Viagra was recently added to Oxford English Dictionary—quite a feat for a relatively new consumer brand.

‡ Chapter 12 addressed certain patient and physician feedback mechanisms *across* indications while these listed reinforcing processes generally take place *within* a given indication.

communication in the form of verbal contact between physicians or published research touting the benefits of a given compound creates a reinforcing feedback process that increases product awareness and willingness to prescribe the new drug.

2. *De facto evidence of efficacy*: Significant adoption of a compound by physicians and its corresponding use by their patients implicitly signals that the product is effective, thereby encouraging other doctors to prescribe it. Berndt and Pindyck of the Massachusetts Institute of Technology (MIT) Program on the Pharmaceutical Industry find, "The fact that a drug is currently used by, say, a million patients is evidence that it is at least somewhat efficacious relative to its side effects and risks" (4). If you have ever tried a particular restaurant after seeing a line of people standing out its entrance, then you have participated in this particular form of reinforcing feedback process.
3. *Standard practice defense*: Fear of lawsuits can affect physician decisions, and prescribing a drug that is a consensus therapy among doctors limits potential exposure to malpractice claims. While such situations may hinder the use of new or innovative compounds, this type of reinforcing feedback mechanism may drive a particular drug to be the "gold standard" of care due to a physician group dynamic designed to reduce their medical liability.

Reinforcing feedback mechanisms can operate at the compound, brand, and/or class level. For example, Berndt and Pindyck analyze reinforcing feedback in the anti-ulcer market at the class level (H_2-antagonists) and the brand level (Tagamet, Axid, Zantac, and Pepcid) (5). The strategic implications of reinforcing feedback mechanisms can be very different if the effects fall at the class as opposed to the brand level. For example, reinforcing feedback related to a specific drug class may imply a strategy of "co-opetition" (6) on the part of companies offering products in that category. Conversely, strong support of a particular brand may be important if the reinforcing mechanisms imply that brand equity is driving the feedback process in an indication marketplace.

These types of reinforcing feedback processes can impact both the ultimate level of sales for a compound and the speed at which those sales be seen (expected sales trajectory), and there are a number of ways to operationalize these feedback mechanisms in the Dynamic Modeling methodology. Word-of-mouth processes affecting physician adoption of compounds were addressed in chapter 8 and need not be revisited here. For simplicity, the second two common types of reinforcing processes can be lumped together under the title of "network effects" and can explicitly be rendered and analyzed using the language of Dynamic Modeling.

CASE STUDY: ILLUSTRATIVE EXAMPLE OF ANALYZING NETWORK EFFECTS USING DYNAMIC MODELING

Imagine a pharmaceutical marketplace with the following characteristics:

1. The indication is a chronic disease in which a number of relatively similar products currently exist.

2. Two pharmaceutical firms are each developing new compounds representing significant advances in therapy which are expected to replace existing therapies upon launch. The compounds are slated to launch in mid-2013 and early 2015, respectively.*
3. Developmental NCEs appear to have identical clinical attributes in terms of efficacy, side effects, tolerability, etc. and will thus be viewed as perfect substitutes for one another.

Further, suppose one of the firms is interested in evaluating the benefits of accelerating the development of its new compound (NCE 2) in order to reach the indication marketplace earlier than the drug being developed by its competitor (NCE 1). For the purposes of illustration, a dynamic model with the following simplifications will be useful to describe this situation:

- Existing therapy options will be grouped into a catchall treatment category called Current (a full mutually exclusive and collectively exhaustive treatment option set of currently available drugs can be included if a more detailed analysis of future patient dynamics regarding treatment changes is warranted).
- Treatment switches are the main determinant in changing patient share (the dynamics of *initiating* and *reinitiating treatment* from the Patient Flow sector can be added as an easy extension).
- Each of the developmental drugs will have a constant and universal Prescriber status of 100% immediately upon release (the full dynamic structure from the Doctor Adoption sector is an easy inclusion).

Figure 2 shows how a modified Standard Template dynamic model can be established to represent this example.

Note that Figure 2 represents a simplified sub-set of structure from the Standard Template as defined in chapter 11, with an addition of some logic in the *Treatment Attractiveness* calculation related to the network effect. In this revised diagram, *utility from network effect* and the *importance of network effect* impact the *Treatment Attractiveness* of the therapy options in the hypothetical indication. In previous incarnations of overall treatment utility, *Treatment Attractiveness* was a function solely of product attributes—efficacy, safety, side effects, etc. (see chap. 10). In effect, the logic shown in Figure 2 allows the utility of the treatment options in an indication to be both a function of specific product attributes and any network effect based on the number of Currently Treated Patients receiving each therapy. Run a mental simulation to see why the structure of Figure 2 represents a reinforcing feedback mechanism. If the *importance of network effect* is nonzero, then *Treatment Attractiveness* for a therapy option will include utility based on the number of patients that treatment has. The more patients on a therapy, the

* The differences in expected launch dates have been exaggerated in this example for illustration purposes only. Detailed information on the relative product profile of NCEs is generally not available that far into the future.

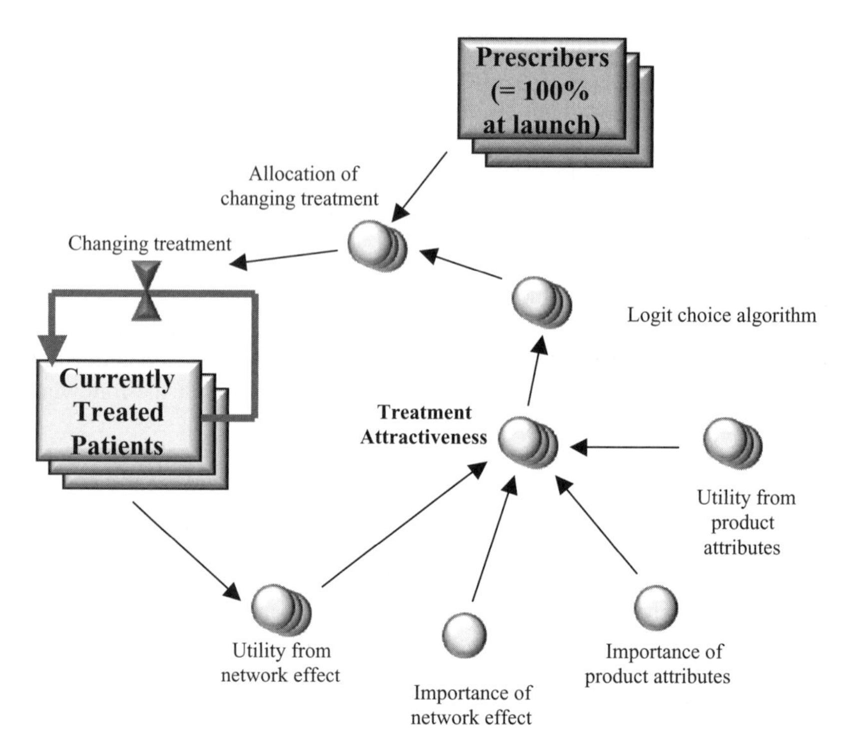

Figure 2 Dynamic Modeling framework for incorporating network effect into Standard Template.

greater the *Treatment Attractiveness* for that treatment, the greater the *allocation of changing treatment* flow to that particular therapy, and so on.

To test the effect of a first-mover advantage, a Dynamic Modeling simulation can be established in which the various strengths of the network effect can be incorporated and the benefits of being first to market can be identified.

USING DYNAMIC MODELING TO ESTIMATE FIRST-MOVER ADVANTAGE

As an initial assumption for this illustrative case study, the dynamic model simulation can first be parameterized to represent a situation in which no network effects are operating in the marketplace. The No Network Effect scenario means that the number of patients receiving each therapy option in the Currently Treated Patients stock **does not** affect the *Treatment Attractiveness* calculation, or by extension the allocation of patients *changing treatment.* Referring back to Figure 2, this scenario establishes that the *importance of network effect* is zero, so that *utility from product attributes* is the only factor in determining *Treatment Attractiveness.* Since in this hypothetical example NCE 1 and NCE 2 are breakthrough therapies

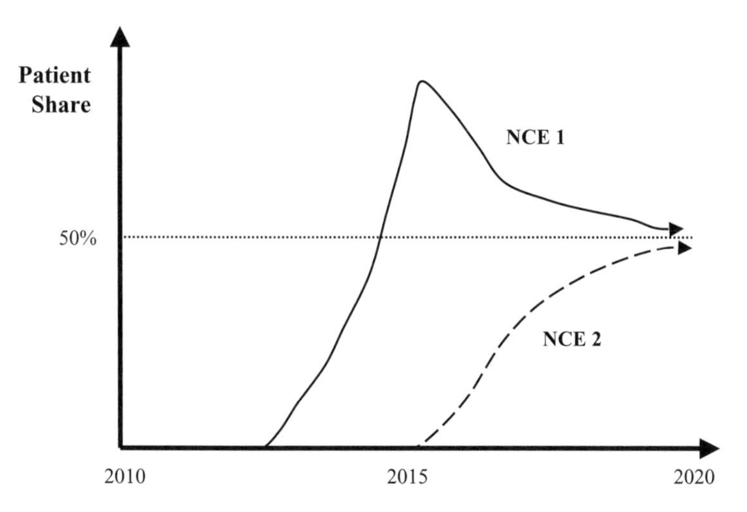

Figure 3 Example dynamic model simulation results under no Network Effect Scenario.

with identical clinical attributes and an assumed 100% Doctor Adoption at launch, the long-run patient shares for each compound should be 50%. As shown in Figure 3, the launch of NCE 1 in 2013 allows it to quickly capture a large share of the market from existing therapies. Until the launch of NCE 2 in 2015, in fact, NCE 1 is establishing a dominant, almost monopoly-like status within the indication marketplace. Since the two compounds have identical attributes and corresponding Treatment Attractiveness, the Logit Choice Algorithm returns an expected 50% share for each (see chap. 11), allowing the two products to eventually divide the marketplace evenly as shown in Figure 3.

In the No Network Effect scenario, the rate at which the initial advantage of NCE 1 is eroded depends on the rates at which patients switch therapies in this simplified example of marketplace dynamics. Upon launch, NCE 2 begins to gain patient share at the expense of NCE 1 as patients switch to the former from the latter. The first-mover advantage for NCE 1 erodes more quickly if the pace of switching between therapies is rapid. Figure 4 shows the results of a dynamic model simulation in which the switching delay is shorter than in the base case. As expected, the first-mover advantage for NCE 1 erodes much more quickly under these conditions, although the long-run equilibrium of 50% patient share for each compound still applies.

Activating the Network Effect

To include the network effect in this example indication, the dynamic model can be reparameterized such that the *importance of network effect* variable takes on some nonzero value. This inclusion will allow the number of Currently Treated

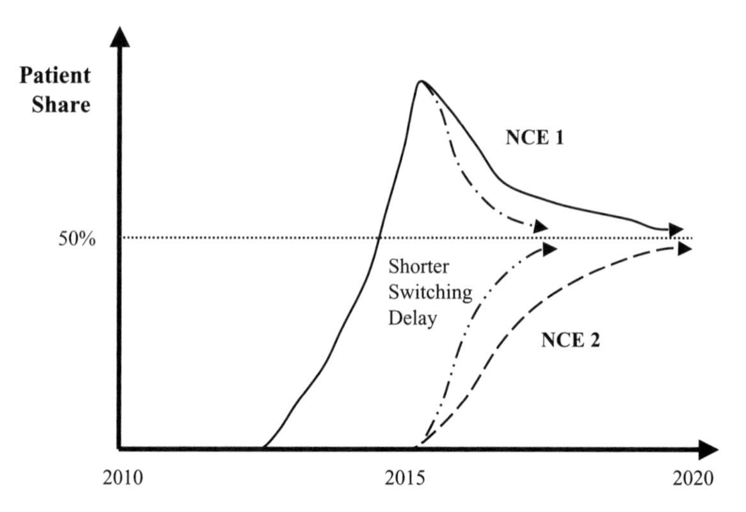

Figure 4 Example dynamic model simulation results under no Network Effect Scenario with shorter switching delays between therapy changes.

Patients to impact the associated *Treatment Attractiveness* of each individual therapy option.* All other conditions in this Network Effect Scenario remain the same as in previous simulation results.

As shown in Figure 5, the Network Effect Scenario causes the first-mover advantage for NCE 1 to persist indefinitely. Although NCE 2 has identical *Treatment Attractiveness* and equal status in terms of Doctor Adoption, it is unable to overcome the initial advantage NCE 1 gains from Currently Treated Patients accumulated during its early post-launch stage.

The results of network effects, such as those shown in Figure 5, build on themselves through a reinforcing feedback mechanism. Accumulated patients on NCE 1 translate into higher utility for that treatment option, thus increasing the share of patients captured and pushing the number of patients receiving the compound ever higher. This network effect locks NCE 1 into a sustainable advantage even though its product attributes and associated Treatment Attractiveness are identical to that of NCE 2.†

In general, the first-mover advantage consists of two components in all cases and a third under circumstances in which network effects are present, summarized in Figure 6.

* The magnitude of this effect can be based on technical analysis such as the study of the ulcer drug market by Berndt and Pindyck but is more often the result of uncertainty analysis and associated team judgment. See chapter 13.

† Network effects can create a sustained advantage even in the case of dissimilar product utilities. Many argue that the Macintosh operating system is inherently superior to its Windows counterpart, but network effects for the latter have resulted in a dominant and persistent market position.

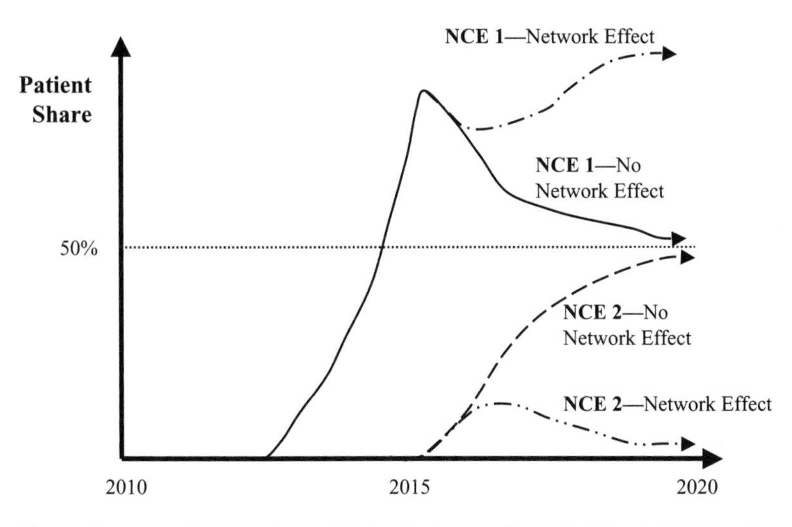

Figure 5 Example dynamic model simulation results under Network Effect Scenario.

1. *Reduced competition period:* Regardless of relative utility evaluations and resulting long-run market shares, the first mover always has an advantage of additional time on the market during which competition is not as severe as other NCEs flounder in development. Market share and profitability during this period depend on the rate at which patients switch between existing therapies and the new NCE.
2. *Transition period:* As new compounds are introduced into the indication, market shares start to shift but have not yet stabilized because of switching delays. During this period, the market is in flux as physicians experiment with different products to determine which is best for individual patients.

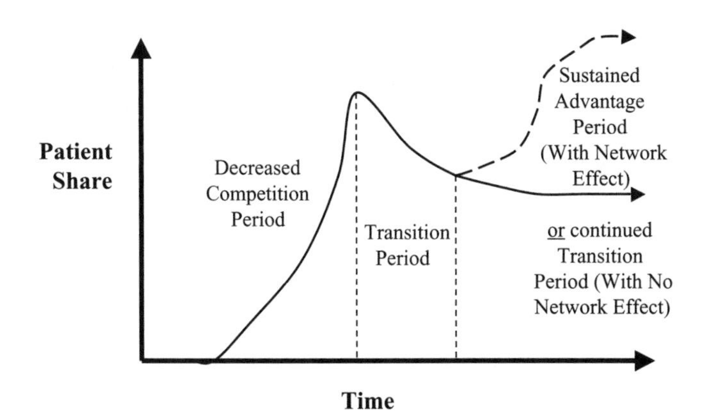

Figure 6 First-mover advantage periods in pharmaceutical markets.

	Match Competition (NCE 2 launch = 2013)	Accelerate Development (NCE 2 launch = 2012)	Most Likely Scenario (NCE 2 launch = 2015)
No Network Effect	No advantage for either compound.	Long-run share for each NCE is 50%, but NCE 2 gains additional market time and revenue.	NCE 1 gains initial lead; long-run share for each compound is 50%.
Network Effect	No advantage for either compound.	Earlier NCE 2 launch cuts into NCE 1 revenue and long-run advantage.	Sustained long-run advantage for NCE 1.

Figure 7 Example strategic scenario grid for accelerated NCE 2 development.

3. *Sustained advantage period:* The third component of first-mover advantage is the steady-stage gain in market share from the network effect, which causes an initial market share increase to persist when similar or even superior compounds are launched. Early product introduction in marketplaces exhibiting network effects creates a sustainable competitive advantage through a reinforcing feedback mechanism.

Dynamic Modeling provides a unique way to quantify and analyze the tradeoffs between accelerated development and associated incremental revenue in pharmaceutical indication marketplaces, particularly in cases in which network effects may be present. In our continuing example, dynamic model simulation results could be used to determine the cost/benefit tradeoffs for NCE 2 as it considers strategies associated with accelerated development. These types of analyses are often detailed and involve financial metrics beyond the scope of this simple example, but can be summarized as shown in Figure 7.

The scenario possibilities embedded in Figure 7 are generally supported with detailed revenue implications determined from the associated dynamic model, as well as cost estimates for various rates of accelerated development of the compound. Depending on the strategic vision of the firm developing NCE 2 and the associated financial analysis from dynamic model simulations, the cost/benefit tradeoffs for compound development times can be reasonably assessed.

SUMMARIZING NETWORK EFFECTS AND FIRST-MOVER ADVANTAGE

Evaluating the incremental costs and benefits of accelerated NCE development are vital in determining the best pace for developing a compound, and inclusion of network effects makes this analysis much more robust. Dynamic model simulations can be used to measure the financial impact of various development strategies under different scenarios, including the strength of the network effect,

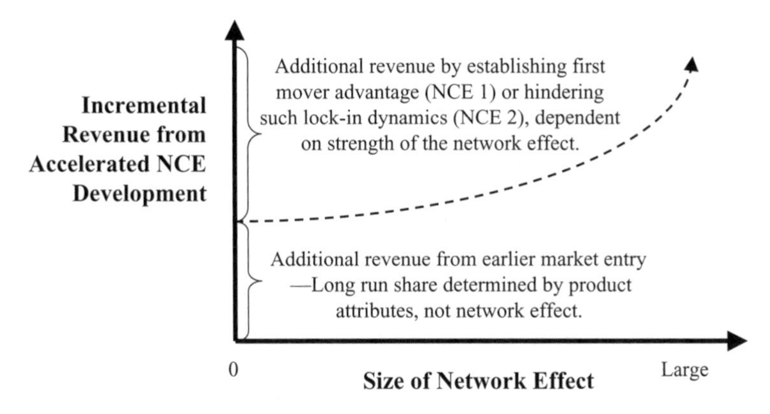

Figure 8 Relationship between network effect and first-mover advantage.

the speed of competitor introduction, and the rate of treatment switching, and the indication marketplace into which compounds will be introduced. In general, however, the first-mover advantages depend on the size of the network effect, as shown in Figure 8.

This chapter has established the strategic importance of reinforcing feedback processes in the form of network effects and has shown how dynamic model simulations can be used to asses the impact of various NCE development decisions. Questions remain, however, as to how the magnitude of the network effect can be estimated and how uncertainty in those measurements can be addressed. Although the problems of data availability, measurement bias, and parameter uncertainty affect all model-building exercises, they are especially severe in the case of completely new therapeutic areas with no existing marketplace dynamics or historical analogs. The next chapter provides some tools for dealing with uncertainty in the extreme case of completely new pharmaceutical markets—techniques that can be applied to issues seen in this chapter regarding parameter estimation of network effects.

SUMMARY

- Feedback mechanisms are sometimes seen in pharmaceutical marketplace and can establish treatment standards based on reinforcing processes.
- These reinforcing feedback mechanisms are often grouped under the heading of network effects.
- The presence of network effects can establish a sustained first-mover advantage that can be decomposed into three distinct phases, reduced competition period; transition period; and sustained advantage period.
- First-mover advantages can be captured using an extension of the Standard Template dynamic model.
- Dynamic Modeling simulations can quantify and analyze first-mover advantages, with or without the presence of network effects in a disease marketplace.

REFERENCES

1. Ding M, Eliashberg J. Structuring the New Product Development Pipeline, Management Science, March 2002.
2. Berndt E, Pindyck R. Network Effects and Diffusion in Pharmaceutical Markets: Antiulcer Drugs. NBER working paper No. 7024, March 1999.
3. Zoltners A, Hahn M, Park S, et al. Analysis of New Product Diffusion Using a Four-Segment Trial Repeat Model. Marketing Science, Summer 1994.
4. Berndt E, Pindyck R. Network Effects and Diffusion in Pharmaceutical Markets: Antiulcer Drugs. NBER working paper No. 7024, March 1999.
5. Berndt E, Pindyck R. Network Effects and Diffusion in Pharmaceutical Markets: Antiulcer Drugs. NBER working paper No. 7024, March 1999.
6. Brandenburger A, Nalebuff B. Co-opetition: A Revolution Mindset That Combines Competition and Cooperation, The Game Theory Strategy That's Changing the Game of Business. Doubleday, 1997.
7. Arthur WB. Increasing Returns and Path Dependence in the Economy. Ann Arbor, MI: University of Michigan Press, 1994.

14

Dynamic Modeling of New Marketplaces: Techniques for Dealing with Uncertainty

INTRODUCTION

This chapter will cover the following topics:

- Uncertainty in new markets and models thereof
- Bounding uncertainty with stock/flow structure
- Reducing uncertainty with
  - sensitivity analysis and
  - maximum entropy estimation.
- Analyzing uncertainty with
  - optimization,
  - real options approaches, and
  - robust decision-making analysis.

Previous chapters have demonstrated the value of Dynamic Modeling in analyzing the introduction of new compounds into existing markets. These models used data from currently available products in existing indication marketplaces to estimate market potential, product attribute utilities, patient treatment dynamics, physician acceptance of new compounds, and other important interrelationships driving market behavior. Although uncertainty about the magnitudes of causal relationships always exists, historical data usually provides solid benchmarks for the parameter estimates that serve as inputs to dynamic model simulations. For example, the hypertension indication is a rather crowded marketplace with well-defined market segments, established evaluations of product attributes, and a wealth of data on past New Chemical Entity (NCE) introductions. These historical analogs can provide a means to analyze historical switching behavior, acceptance of newly released products, and the overall pace of new product diffusion. A dynamic model built for the purpose of developing strategies in the hypertension market could therefore be built on a solid foundation of historical experience and indication data.

Problems arise, however, when historical data does not exist or is too old to be strategically relevant. In this chapter, we consider the case of novel compounds in new or undeveloped markets and show how the Dynamic Modeling process can be used effectively in situations of high uncertainty. In such cases, traditional forecasting or statistical methodologies become impractical as there in no past data from which future trends can be extrapolated. In these types of situations, the operational nature of Dynamic Modeling is even more crucial to strategic success.

CASE STUDY: USING DYNAMIC MODELING IN LATENT MARKETPLACES

Several years ago, one of our very first clients was developing a compound for an indication in which only one treatment had existed for the last quarter century. The existing therapy was viewed as an ineffective treatment for the disease, but was the only game in town in terms of pharmaceutical intervention. Clearly, this market had no relevant information about new compound introductions; no analogous product to which the NCE could be compared; and only a very shaky estimate of market size. Despite the lack of data, the dynamic model proved very useful by operationally defining the strategic issues facing the introduction of the compound. By providing a means to test potential strategies, the client team was able to evaluate the relative impact of alternative marketing and brand positioning approaches, assess the importance of various model inputs on the commercial evaluation of the compound, and prioritize the supporting data collection process. The resulting Dynamic Modeling simulation was an integral part of the marketing team's Brand Planning related to the eventual successful launch of the NCE.

THE ROLE OF UNCERTAINTY IN MENTAL AND DYNAMIC MODELS

Before diving into the techniques for coping with uncertainty, it is helpful to set a useful framework for assessing the value of Dynamic Modeling. It is tempting to evaluate models against the unrealistic standard of the model that would be built if all the required data inputs were available from controlled experiments. But the alternatives to explicit simulation models are mental models that have been shown to suffer from serious problems of consistency, limited processing capacity, and short memory (1,2). The problem of data availability exists for both mental and simulation models, and the only relevant question is how to make the best use of the information that is available.

As was discussed in chapters 1 and 2, Dynamic Modeling is a useful way to make mental models explicit and testable in a concise and rigorous manner. Our cumulative experience over many years of building dynamic models is that augmenting mental models with simulation improves the quality of analysis and results in better strategic decisions (3). Mental models provide creative insights that no computerized technique can match, and combining them with simulation can provide increased clarity and consistency in the decision-making process. In successful Dynamic Modeling projects, the process of building a simulation model synergistically improves the quality of the mental model that, in turn, produces an improved simulation model from which better strategic decisions can be made.

BOUNDING THE POSSIBILITIES: STOCK/FLOW STRUCTURE

One problem in strategic analysis of very new markets is that the possibilities for marketplace evolution are almost infinite. For example, the Case for the market could be very big or very small, the Place of the new compound might be affected by new market entrants or changing treatment paradigms, and the Pace of the NCE market uptake could be extremely rapid or excruciatingly slow (see chap. 12). Financial analysts face the reality of unbounded possibilities when they try to draw the expected sales trajectory (see chap. 1) of a compound in a very new indication marketplace, as shown in Figure 1.

The questions posed in Figure 1 and the range of possible answers make it vitally important to limit the possible outcomes for a compound based on the strategic decisions made by the firm. We would like to say confidently that a set of actions will result in sales that fall in a particular range and not in some other range—a process called **bounding**.

The first step in bounding potential outcomes is to specify an operational model structure consisting of a series of interconnected stocks and flow. Previous chapters have provided many examples of the stock/flow physics found in most indication marketplaces; epidemiology concepts such as incidence, prevalence, and mortality, patient dynamics of switching, discontinuation, and reinitiation of treatment, and physician dynamics involving awareness, trial, and usage of pharmaceutical products. Many of these structures must also apply to new markets, although data to populate them may be nonexistent. For example, the fundamental

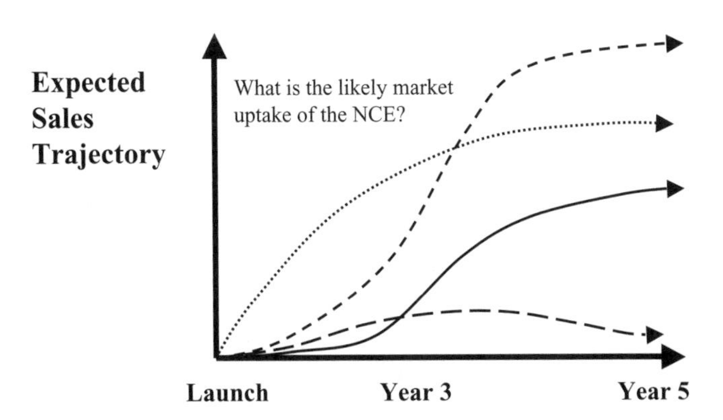

Figure 1 How can the expected sales trajectory of an NCE in a *new* indication marketplace be evaluated?

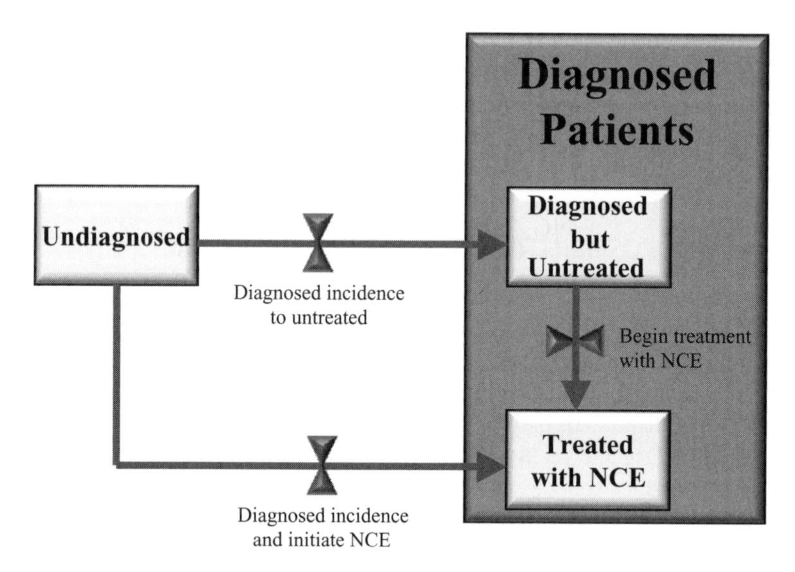

Figure 2 Stock/flow structure of Patient Flow dynamics in new pharmaceutical markets.

"physics" of epidemiology dictate that patients will eventually flow from the undiagnosed state to the treated state—if they survive that long. Other patients may have received a disease diagnosis but have not been treated with prescription medication as none had previously existed, as shown in Figure 2.

Including the Undiagnosed and Diagnosed, but not Untreated stocks, in a dynamic model creates an explicit delay between the introduction of an NCE and the initiation of treatment by patients in a truly new marketplace. This fundamental result is not open to debate and is not a function of parameter estimates—it is an unavoidable principle of latent indication marketplaces. Bounding the possibilities with operational stock/flow structure not only places some limits on the speed at which a new compound might be able to capture patients, but also demands a causal explanation of what will drive patients to move from their previously untreated states to the NCE.

The best source of information for the stock/flow diagram of new marketplace dynamics is often the Brand Planning team itself. They have likely been studying the indication for some time, and team members may have experience in other markets that proves valuable. Our experience with clients is that the stock/flow concept is easy to understand and one or two facilitated workshops allow the team to create the basic Dynamic Modeling diagram. An experienced modeler can suggest model structures that have worked for other indications such as those detailed in previous chapters, and team members can modify these templates to fit the particular case. The process of creating a consensus stock/flow map leverages the experience of a dynamic modeler and the knowledge of various members of the Brand Planning team. Doing so, places at least some reasonable limits on the ultimate sales trajectory of a new compound as the structure/behavior paradigm (see chap. 2) of Dynamic Modeling dictates that the behavior of the system will ultimately be a function of its stock/flow structure and associated data inputs.

APPROACHES FOR COPING WITH UNCERTAINTY IN MODEL RELATIONSHIPS

Our experience with clients in the pharmaceutical industry is that the combination of two approaches can effectively address inherent uncertainty in the data necessary to populate a dynamic model. Not all modeling efforts require the techniques that follow, but they do offer recourse in cases of limited data availability such as the introduction of a compound into a truly new market. The two approaches/techniques are as follows:

1. *Reduce uncertainty by better measurement*: Use a combination of team judgment and historical data to estimate model inputs as precisely as possible. Sensitivity analysis of Dynamic Modeling results can then be used to allocate resources to the most important/sensitive model parameters. This process helps identify which model parameters have the most impact on overall model results, and focuses the effort on getting those parameter estimates as "right" as possible.
2. *Use analytic techniques to embrace uncertainty*: Acknowledge the inherent uncertainty around new product launches and use modeling to account for that uncertainty in a rigorous fashion. Techniques for managing uncertainty include real options, optimizing resource allocation under uncertainty, and computerized search for robust strategies. Modeling is a prerequisite for all techniques of managing uncertainty, and the dynamic models developed in previous chapters can be readily extended to include these approaches.

Using Sensitivity Analysis to Determine Dynamic Modeling Parameters

Model parameterization involves an iterative process in which the model itself is used to determine the parameters that need precise estimation. As discussed in previous sections, the first step in model building is to create the stock/flow structure. Once the stock/flow structure is complete, the relationships that control the flows must be specified and those relationships always involve unknown parameters. A helpful technique is to make rough estimates of the parameters and use the model to run sensitivity analysis, which determines the parameters that have the largest impact on important model metrics such as market share and profitability. An example of this type of sensitivity analysis is a tornado chart, as shown in Figure 2, which shows the effect of various parameter on a theoretical model metric.

Tornado charts such as the one in Figure 3 help establish the importance of various input data on overall model results. Effort should be allocated to estimating

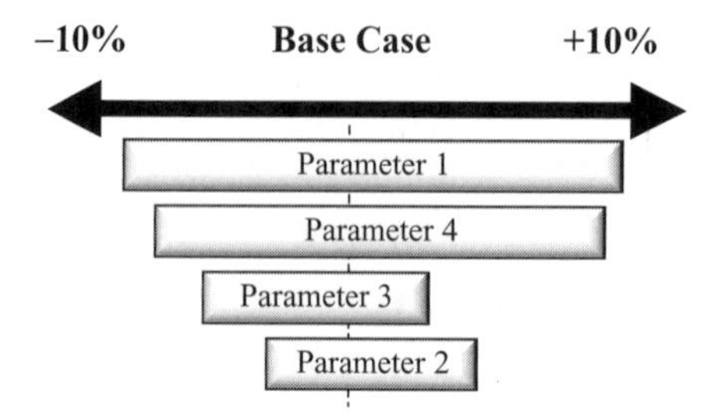

Figure 3 Example tornado chart detailing impact of increases and decreases of various parameter values on dynamic model output metric.

the important parameters more precisely while estimates of unimportant parameters, which have less of an impact, need not be changed.

For example, consider the new product model in chapter 13 that was used to demonstrate the strengths of network effects in a given indication. Sensitivity analysis determined that there were three important parameters in the model: the long-term share taken by the new pharmaceutical class, the pace of switching between developmental NCEs and existing products, and the magnitude of the network effect in determining treatment attractiveness. Each of these parameters can be estimated in a prelaunch situation when there is no relevant historical data available.

Long-run share of the new pharmaceutical category:

- First, the most important technique for estimating the long-term share is a market research study that involves the analysis of physician preferences. Chapter 10 detailed how physician evaluations of treatment attributes can be used to estimate the attractiveness of existing products and new compounds still in development. Chapter 11 then demonstrated how these attractiveness ratings can be used in various Logit Choice algorithms to determine share calculations in a dynamic model.
- Second, the results of such conjoint surveys should be combined with managerial judgment. Brand Planning teams focusing on new compound introduction usually include clinicians who are experienced with existing therapies in an indication. Techniques from group model building and decision analysis can be used to systematize the process of eliciting reasonable ranges for parameter estimates from panels of experts.
- Third, studies of past product introductions can be used to estimate market potential. For example, a study by Zoltners et al. (4) examined a sample of 21 pharmaceutical products to estimate how new product characteristics affect repeat purchasing and product diffusion. The data required to analyze analogous new product introductions are available from internal pharmaceutical companies records, vendors such as IMS Health, and increasingly other third-party data providers such as those mentioned in chapter 6. The ready availability of historical data makes it possible to

create a database of parameter values that can be applied to new product introductions in different indications.

Pace of switching between the new compounds: Market research studies are not particularly useful for studying dynamic switching behavior, making it necessary to rely on historical data and detailed team knowledge of the switching process. Chapter 6 explained how longitudinal claims databases are used to estimate the probability of switching from one treatment to another in a defined indication. In addition, the Zoltners' study (4) described in the previous section measures the impact of various factors on the pace of treatment switching. Finally, qualitative information can be very helpful in measuring switching delays. In some indications, patients see their physicians frequently to check the effectiveness of their treatment. In these disease markets, the pace of switching is often rapid while treatment changes can take much longer in indications where existing treatments are satisfactory and continuous monitoring is not required. Clinical experts can often provide qualitative information on the relative frequency of physician interaction in a particular disease marketplace.

Strength of the network effect: Regression-type analysis can be used to estimate the relationship between the patient base receiving a given treatment and the probability that additional patients will adopt or switch to that compound (5). The network effect parameters derived from statistical analysis of past therapies can be used to set ranges for the parameters related to the new product introduction.

The magnitude of the network effect can also be estimated based on expert judgment. Our experience is that the network effect can be best interpreted as an elasticity with respect to therapy utility that ranges from 0 to 0.7. An elasticity of 0.2 means that a 10% increase in the number of patients receiving a particular therapy causes a 2% increase in corresponding Treatment Attractiveness for that compound. The network effect is likely to fall in the upper part of the range when the following occur:

- There is a strong communication network between physicians about the effectiveness of a therapy through consultation, conferences, or continuing medical education.
- There is a core group of key opinion leaders (KOLs) who strongly influence the prescribing decisions of other practicing physicians.
- Insurance plans authorize the use of products based on what doctors prescribe most often.
- Pharmaceutical companies promote physician sharing of information about product effectiveness.

Evaluating initial estimates for these types of parameters through sensitivity analysis can be used to test both their importance on overall marketplace behavior and establishes ranges that can be used in other analytic techniques.

Analyzing Potential Market Outcomes to Determine Dynamic Modeling Parameters

Model parameters can also be calibrated by combining prior ranges for parameter values with estimated ranges of market outcomes. For example, assume that through a combination of judgment and historical examples the Brand Planning team has arrived at ranges for the parameter values to serve as inputs to a dynamic model. In addition, the team believes that it can set ranges for the expected sales trajectory of the compound and perhaps other key market metrics. In many cases, it is easier to estimate a range for potential market behavior than it is to directly estimate the parameter values which drive that behavior. The ranges are often stated in the form, "In two years market share for the NCE will be no less than 2% and no more than 15%." Similarly, conditional ranges can be expressed in the form, "If cumulative marketing spending is a least $20 million, market share for the compound will be a least 10%."

The estimated ranges the compound's expected sales trajectory can be combined with initial parameter estimates to generate new parameters that are consistent with all the information specified about the new compound. A technique called **maximum entropy estimation** can be used to find the set of parameters that satisfies the range constraints for market outcomes and also keeps these new input estimates close to their initial values (6). Maximum entropy estimation extracts the maximum amount of information about the model parameter values from the assumptions made about the ranges of potential future outcomes.

MANAGING UNCERTAINTY AROUND DYNAMIC MODELING PARAMETERS

While the previous section discussed techniques for quantifying the magnitude of the causal relationships that drive subsequent market behavior in an indication, substantial uncertainty will remain no matter how much skill and effort is applied to estimating the important parameters accurately. In our client engagements, we have found three techniques to be extremely useful in managing uncertainty around the process of developing and launching new pharmaceutical products. The following three subsections list and explain these techniques.

Optimization Under Uncertainty

Optimization under uncertainty searches for the set of decisions that maximizes an objective; such as Net Present Value (NPV) of a compound, given that there is uncertainty about the underlying model (7). The objective can be more sophisticated than a single point value, such as the average NPV across multiple scenarios, the results under the constraint that NPV cannot fall below a certain level, or the percent of outcomes that land above a given threshold. Alternative simulations are created by drawing parameter values from probability distributions and then evaluating the objective function for multiple simulations. This Monte Carlo

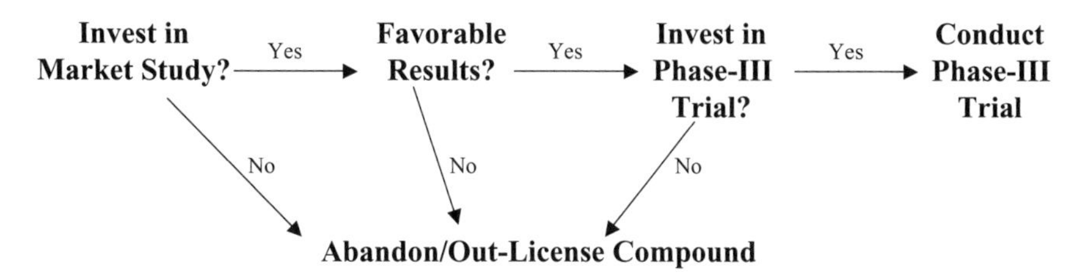

Figure 4 Decomposing a Phase III clinical trial decision using a real options analysis.

approach generates a distribution of potential outcomes based on alternative strategies.*

Optimization under uncertainty helps avoid the trap of developing a strategy that is only effective under a very specific set of market assumptions. Accounting for uncertainty results in strategies that are more robust in the sense that they are reasonably effective under many different market assumptions. Many of our clients find that robustness can be increased by setting stricter boundaries on the value of the objective function such as NPV. The optimized strategies resulting from this type of uncertainty analysis are more robust in the face of an uncertain future and lack of data on the parameters driving marketplace behavior.

Real Options Analysis

Real options analysis exploits uncertainty by taking advantage of favorable outcomes and limiting the cost of unfavorable outcomes. Real options analysis can create value by more effectively allocating resources between competing projects. The technique breaks large decisions into sequences of smaller components and considers the distributions of outcomes that could be caused by each sequential choice (8).

Case Study: Real Options Related to Phase III Clinical Trials

Consider the decision to enter an NCE into Phase III clinical trials. The commercial assessment of the compound is at best a rough estimate but the uncertainty can be reduced significantly with a market survey or conjoint analysis study showing how the NCE might be perceived in the marketplace. The market survey is relatively inexpensive while phase III trials are very expensive and time consuming. The real options approach decomposes the investment decision into two parts: the decision to invest in the market study and the decision to invest in Phase III trials, as shown in Figure 4.

Investing in the market study buys an option, similar to a call option on a stock, to start Phase III trials if the market study is favorable. The purchase price of the option is the cost of the market study, the exercise price is the cost of the Phase III trials, and the payoff is the expected value of the compound conditional

* See chapter 16 for a description of Monte Carlo related to portfolio planning.

on positive results from a market survey. The NCE can be abandoned/out-licensed and Phase III costs avoided if the market study is unfavorable. Depending on the specific costs and the compound's commercial assessments, it is possible that the option is valuable and the firm should proceed with the market study. Instead of making an "all or nothing" bet, the real option approach has allowed the team to make a smaller decision, the results of which will be instrumental in making the larger decision on the basis of better information.

Real options analysis can be combined with the Dynamic Modeling approach described in previous chapters. To assign values to real options, the model must include decision rules that trigger investment decisions when certain conditions are met. In this example case study, Phase III trials would begin if the results of market survey showed the NCE had a significant chance to earn significant revenue, over and above the incremental costs (including the clinical trial) of bringing it to market. The form of the real options decision rule is usually determined through the collective wisdom of the Brand Planning team, but optimization algorithms and Monte Carlo simulation results also can be employed to develop a more informed set of decision criteria (9).

Robust Decision Making

Robust decision making is a set of cutting-edge techniques developed by researchers from the Rand Corporation for analysis of situations involving deep uncertainty (10). Robustness is an important decision criterion because it addresses the risk of very bad outcomes. Deep uncertainty refers to situations where there is little agreement about objectives, the underlying model, or the distributions of parameter values. Research has shown that exploratory modeling techniques using the robustness criterion can be useful in conditions of deep uncertainty.

Exploratory modeling usually involves the creation of relatively simple dynamic models that are similar to the ones described in previous chapters. Multiple simulations are run over a wide range of parameter values to generate a collection of possible model outcomes. The input parameters are not assumed to come from probability distributions are there is no claim that some values are more likely than others.

The exploratory modeling simulations are used to test the robustness of different strategies, with robustness is defined using the metric of regret. Regret is the difference between the value of a model output metric in a particular simulation under a set of decisions and the value of that same metric under the optimal decisions for that situation (11). In other words, regret is the difference between what happened versus what would have happened if an optimal set of strategies had been followed. A set of decisions is said to be more robust if it has lower regret than alternative strategies and the most robust decision set results in the lowest regret.

Visualization techniques can be used to analyze the robustness of alternative sets of decisions. For example, the regret for a strategy of rapid development of an NCE could be displayed for alternative values of network effects and the pace of various market dynamics such as Patient Flow and Doctor Adoption. Graphics packages are useful for rendering and visualizing the robustness of strategic decisions under cases of uncertainty.

OVERVIEW

Overall, this chapter has presented two complementary approaches to the uncertainty created by very new markets. The first approach is to specify the model structure and then the parameters as accurately as possible. The second approach is to acknowledge that substantial uncertainty will always exist and to manage the uncertainty as effectively as possible. Optimization under uncertainty, real options, and robust decision making can all be effective in exploiting uncertainty.

SUMMARY

- Modeling of new marketplaces is inherently uncertain due to lack of data and historical analogs.
- Dynamic Modeling can bound the potential outcomes for an NCE introduction into a new market by establishing operational stock/flow structure dictating subsequent marketplace behavior.
- Uncertainty in model parameters can be reduced by applying techniques of sensitivity analysis and maximum entropy estimations.
- Uncertainly in model parameters can be analyzed using optimization techniques, real options approaches, and decision robustness evaluations.

REFERENCES

1. Sterman J. Business Dynamics: Systems Thinking and Modeling for a Complex World Boston, MA: McGraw Hill, 2002.
2. Bonabeau E. Predicting the unpredictable. Harvard Business Review, March 2002.
3. Vennix JAM. Group Model Building. Chichester, UK: John Wiley, 1996.
4. Zoltners A, Hahn M, Park S, et al. Analysis of New Product Diffusion Using a Four-Segment Trial Repeat Model. Marketing Science, Summer 1994.
5. Berndt E, Pindyck R. Network Effects and Diffusion in Pharmaceutical Markets: Antiulcer Drugs. NBER working paper No. 7024, March 1999.
6. Golan A, Judge G, Miller D. Maximum Entropy Econometrics: Robust Estimation with Limited Data. John Wiley, 1996.
7. OptQuest for Crystal Ball 2000, Users Manual, Decisioneering, Inc.
8. Trigeorgis L. Real Options and Business Strategy. Risk Books, 1999.
9. Schwartz E. Evaluating Investments in Disruptive Technologies. UCLA Anderson School of Management, May 2001.
10. Bankes S, Lempert R, Popper S. Shaping the Next One Hundred Years. Rand Pardee Center, 2003.
11. Savage LJ. The Foundation of Statistics. John Wiley, 1950.
12. Homer J. A Dynamic Model for analyzing the emergence of new Medical technologies. PhD thesis, MIT Sloan School of Management, Unpublished.

15

Integrating Standard Template Dynamic Models Across Indications

INTRODUCTION

This chapter will cover the following topics:

- Problems with simple aggregations of indication models
- Comorbidity overlap of patients
- The dynamics associated with misdiagnosis of patients
- Prescriber feedback between indications
- Patient feedback between indications

Previous chapters have shown how the Standard Template is designed to be indication specific and usually addresses the dynamics in a particular existing or potential marketplace. Many of our clients have chosen to create separate and independent Standard Template models for a number of indications of interest, with resulting analysis applying to only those specific indications in isolation. The Dynamic Modeling approach allows for these individual models to be combined into an integrated master or enterprise model that encapsulates all relevant indications, allowing for a firm's overall portfolio positioning to be evaluated. While these aggregations are often done as simple roll-ups in many types of traditional forecasting efforts, the Dynamic Modeling approach has the added benefit of incorporating feedback between indications, either in the form of patient word of mouth, physician prescribing patterns, and/or patient movement between indications over time. Such dynamics are extremely difficult to capture in static-based methodologies, and the level of insight provided by Dynamic Modeling in this arena can be a great source of competitive advantage.

Recall that the Standard Template dynamic model incorporates three main sectors to arrive at an operational and actionable representation factors, which drive change in pharmaceutical marketplaces (Fig. 1 in chap. 2). As was demonstrated

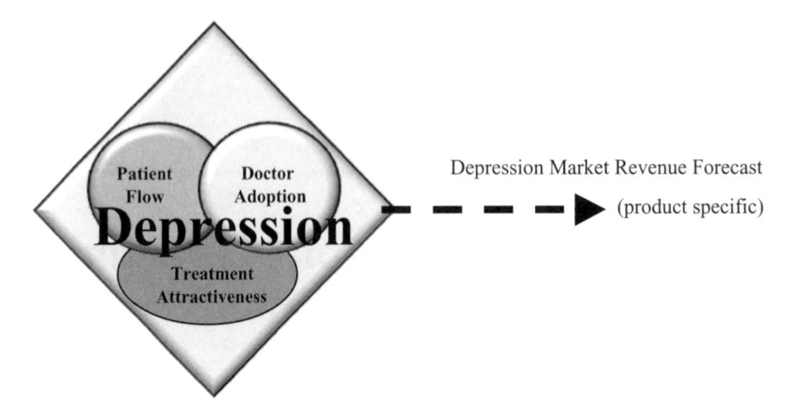

Figure 1 Standard Template for depression resulting in indication forecast.

in earlier chapters, the Standard Template can also translate resulting patient dynamics into a revenue forecast based on parameters such as *true compliance* and *average price of therapy*. In the preceding chapters, such analyses have been indication specific, but a true integrated approach to market positioning demands that the interplay between indications be addressed, and Dynamic Modeling provides the means to do so comprehensively.

For example, a Standard Template model of the Depression market might be used as another methodology for commercial assessment of the compound and as a tool for developing strategies within the indication, as shown in Figure 1.

Since most pharmaceutical firms have drug offerings in more than one indication, the question becomes how to account for multiple therapeutic areas or disease categories within a single modeling framework. At first glance, it appears possible to simply aggregate (in isolation) the Standard Template results for various indications to arrive at a revenue projection for the industry as a whole. Conceptually, such a situation might look like Figure 2.

At the very least, a series of Standard Template dynamic models can be based on a common platform of consistent population and epidemiology numbers. But a simple "adding up the indications" as depicted in Figure 2 may or may not be reflective of a few key portfolio considerations:

1. *Overlap of patients*—Congestive heart failure (CHF) patients may have a hypertension (HTN) comorbidity, for example. Double counting such patients would overstate the expected sales trajectory for a drug that is used to treat both indications.
2. *Misdiagnosis*—Some patients may be incorrectly diagnosed and treated for one disease, but should actually be accounted for in another indication. Patient dynamics may show a movement over time of patients to a correct diagnosis,

Figure 2 Additive representation of Standard Template results for various indication marketplaces.

and it is crucial to capture such reallocation across indications to ensure consistent methodologies and assumptions in various indication evaluations.

3. *Physician prescribing feedback*—Physicians may be more willing to prescribe a drug in a new indication after having positive experiences with it in earlier indications. This reinforcing feedback effect is explained within an indication in Chapter 8, but here refers to cross-indication feedback.
4. *Patient word of mouth*—Treated patients may be purveyors of good news, encouraging others to seek treatment for particular indications. The resulting increased rate at which patients interface with the health care system, combined with the wide availability of health-related information via the Internet, makes it more likely that patients will be diagnosed for other medical issues they may be experiencing. Such word-of-mouth effects could therefore influence not only diagnosis and treatment rates in a given indication, but those in other diseases as well.

Each of these possibilities will be addressed as we extend the Dynamic Modeling approach into the realm of multiple indications with an existing portfolio.

OVERLAP OF PATIENTS

Comorbidity issues often plague traditional market evaluation efforts, as accounting for cross-indication overlap can be exceedingly difficult. For example, the picture of separate indications for CHF and HTN may in fact contain patients suffering from both diseases simultaneously. In such situations, simply combining indication-specific models of these markets may overstate the forecasted potential of drugs intending to treat both indications, as shown in Figure 3.

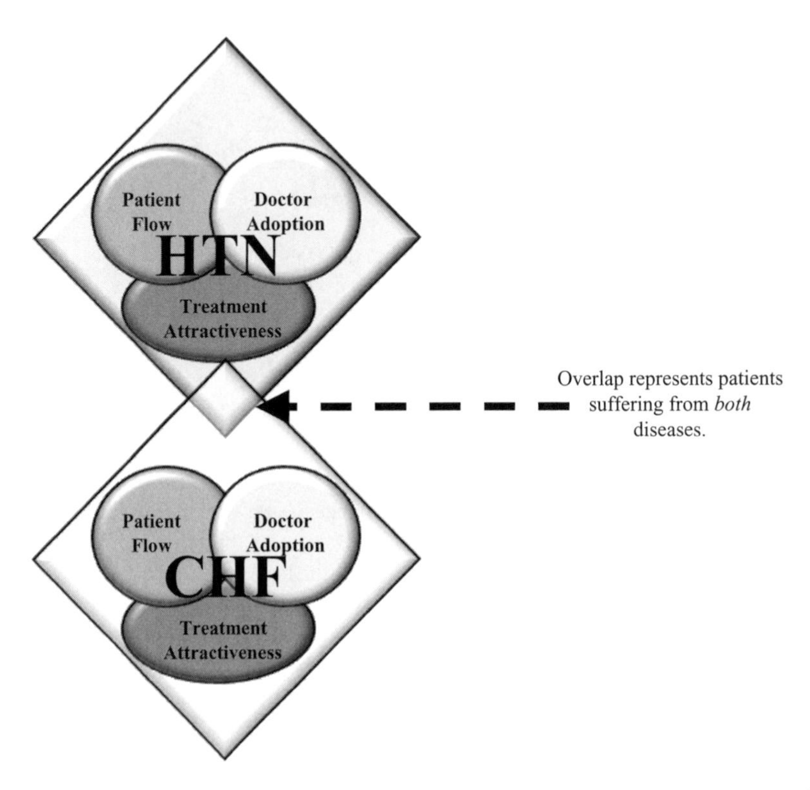

Figure 3 Depiction of overlap between hypertension (HTN) and congestive heart failure (CHF) patients.

Dynamic Modeling on a portfolio level allows this overlap to be explicitly addressed in one of two very simple ways*:

1. *Comorbidity disaggregation*—As we mentioned in chapter 5, the Standard Template allows for patients to be disaggregated into specific segments of strategic interest. In the case of indication overlap, comorbidities are often a useful choice for such aggregations. Continuing our example, HTN patients could be disaggregated into the followings elements:
   a. HTN with no comorbidities
   b. HTN with CHF comorbidity

Patients in category (b) would then be taken out of the CHF Standard Template model to avoid double counting. Conversely, this comorbidity disaggregation could reside within the CHF model, with patients in the element of CHF with HTN being extracted from the corresponding HTN model. In either case, establishing

* Another approach, using an agent-based modeling (ABM) framework, was addressed in chapter 9.

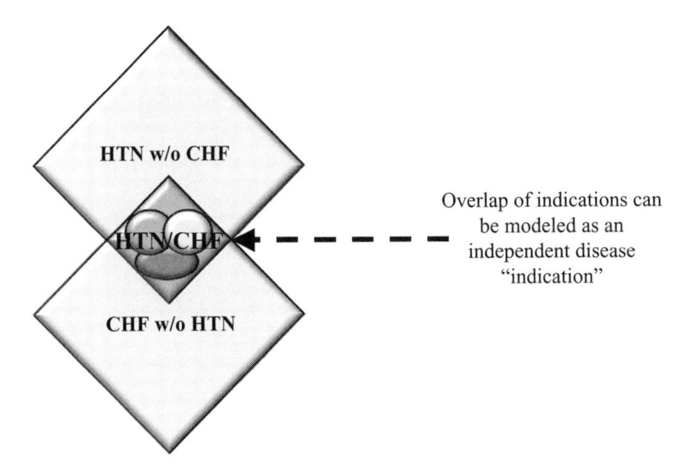

Figure 4 Overlap of HTN and CHF indications being treated as a separate indication.

relevant comorbidity disaggregations with clearly defined MECE elements (see chap. 5) is vital to correct quantification of comorbid patient segments.

2. *Separate CHF/HTN Standard Template model*—If there is a significant overlap of indications and the possibility of differential dynamics within the intersection, a combined indication Standard Template can be developed. In this case, a separate dynamic model can be created that includes CHF/HTN patients as an entirely new indication all to itself, as shown in Figure 4.

Either option dealing with the overlap of patients explicitly accounts for patients that might otherwise be included in multiple indications, creating the possibility for inflated projections due to double counting. The concept of an MECE set with regards to comorbidity ensures that patients are captured once, and only once, in each potential patient segmentation along indication lines.

MISDIAGNOSIS

Recall from our description of the extended Standard Template the possibility of including misdiagnosis as a distinct subset of the stock of Undiagnosed Patients (see chap. 5.) This approach works fine in "one off" indication models, but has the potential to break down when looking at combinations of indications at the portfolio level. Bipolar Disorder patients, for example, are often misdiagnosed as having Depression. In such a situation, the dynamics that might play out between these two indications are shown in Figure 5.

Clearly, Bipolar patients who have been misdiagnosed as having Depression exist in the Standard Template depictions of **both** markets. This situation is not merely a question of overlap, as in the previous example. In this case, there is an actual dynamic flow of *corrected incidence* that would take patients from the

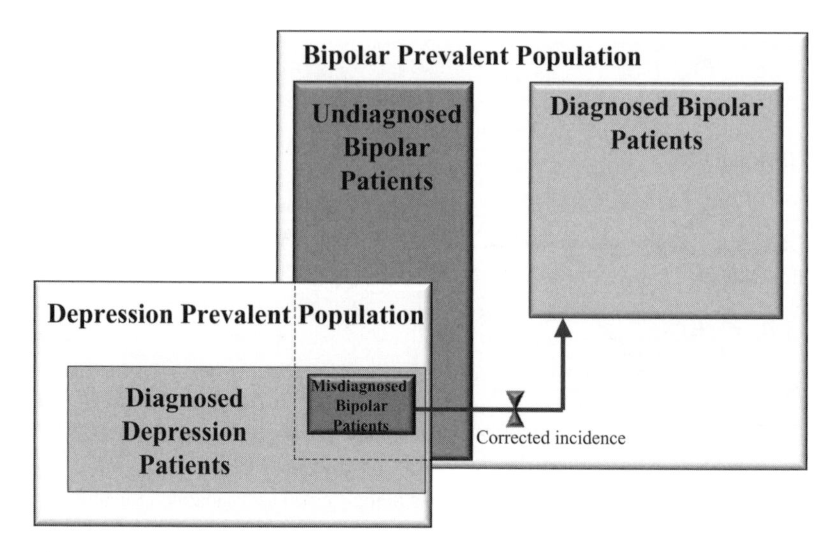

Figure 5 Example dynamics of misdiagnosis and *corrected incidence* between Depression and Bipolar Indications.

stock of Misdiagnosed Bipolar Patients and send them into the Diagnosed Bipolar Patients category. In effect, these patients should now be captured in the Bipolar dynamic model, and no longer exist in the Depression category. Allowing for this possibility in a dynamic model that encompasses multiple disease categories gives both a clearer picture of the movements of patients between indications and ensures the correct accounting for patients as they progress from being improperly diagnosed to receiving a correct diagnosis.

Strategic Implications of Combining Dynamic Models Across Indications

Standard Template dynamic models are designed to be indication specific, providing a way to apply patient-level longitudinal data to aggregate disease epidemiology numbers in order to calculate actual Patient Flow among treatment states. In addition, adoption of pharmaceutical products within a given therapeutic area is a key part of the Doctor Adoption component. But there are often key interrelationships among disease marketplaces that from a portfolio perspective require an integrated approach to analyzing multiple indications.

Comorbidities are the rule rather than the exception in a number of key therapeutic areas, particularly due to aging of the population and increasing diagnosis rates of individual diseases, but the existence of multiple diseases within a target population often causes problems in traditional marketplace evaluations in three key ways:

1. Compounds used to treat multiple diseases must guard against double counting of potential patients when analyzing epidemiology projections.
2. Overlap of patients across comorbid conditions can also present challenges to effective patient segmentation.
3. Misdiagnosis dynamics can introduce challenges in evaluating existing and/or potential marketplaces for a compound.

As such, an integrated framework for establishing evaluations of therapeutic areas, alone and in combination, is essential for analyzing key interrelationships between indications. Portfolio management demands that such a framework, such as the one Dynamic Modeling provides, is able to capture these interactions in a consistent and manageable fashion.

For example, one of our clients was developing a Brand Plan for a compound whose primary indication was a major chronic disease, but a secondary indication in a related illness was being considered as well. The following issues surrounding these often comorbid conditions were difficult to address:

- How big was the comorbid population, and how did those estimates vary by patient segment?
- How did existing drugs in each marketplace fare in terms of single versus multiple indications?
- How frequently did patients in the primary indication develop the disease of the secondary indication, and vice versa?
- How would the strategy vary depending on pursuing a single versus multiple indications?

Standard Template dynamic models had been developed for each potential indication in isolation, but an integrated indication model was necessary in order to answer some of these key strategic questions. By creating three models that ran simultaneously—one for the primary indication only, one for the secondary indication only, and one for patients suffering from both diseases—the Brand Planning team was able to evaluate the potential interaction effects between disease marketplaces, including the potential for physician prescribing feedback that might accelerate adoption in the secondary indication. The team was able to use the dynamic model to test a number of potential strategies and arrive at an effective set of positioning and associated marketing initiatives related to the compound.

Integrating indication-specific Standard Template models allows fundamental interactions between disease marketplaces and therapeutic areas to be explicitly addressed, providing a much clearer and more comprehensive picture of a firm's overall portfolio position. Such combinations are an added benefit of the Dynamic Modeling framework, as its flexibility allows for such adaptations to be made in a rigorous and consistent fashion.

PHYSICIAN PRESCRIBING FEEDBACK

Pharmacologic treatments often have multiple indications for which they are eventually prescribed, but such designations do not take place simultaneously. In fact, individual drugs often add secondary and tertiary indications over time as the effects and usage of their primary indications are observed in the marketplace. Off-label usage, though not encouraged, often becomes more and more accepted over the course of a drug's patent life as physicians and patients alike become more comfortable with the demonstrated safety of a particular treatment.

In one of our client projects, members of the working team were convinced that initial indications saw much slower acceptance on the part of physicians than subsequent indications did. There were a number of hypotheses for such phenomena: brand awareness, relationship with sales representatives, proof of efficacy in initial indications, etc. Regardless of the reasons, the client wanted feedback dynamics to be included at the portfolio level to assess the impact of multiple indications on the Doctor Adoption sector of the Standard Template model. For example, Prescribing Physicians for Indication 1 might influence other doctors, both within the indication and in subsequent indications, through a reinforcing word-of-mouth process, as shown in Figure 6.

The feedback principles shown in Figure 6 are very easy to include in the Doctor Adoption structure (see chap. 8), and are especially powerful when creating models at the portfolio level. Allowing for physician word of mouth to impact

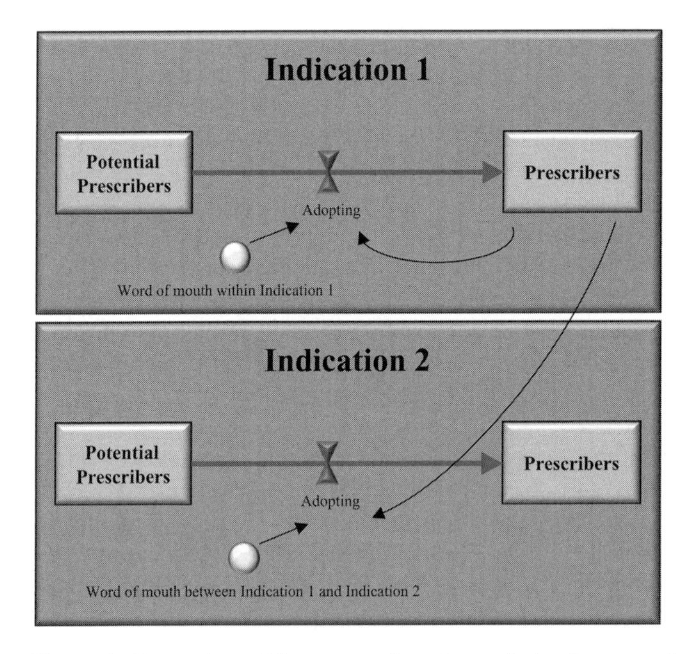

Figure 6 Dynamic model representation of cross-indication prescriber feedback effects.

multiple indications and quantifying the strength of those effects using the Dynamic Modeling frameworks provided our client with a more realistic depiction of actual marketplace dynamics. It also allowed them to quantify the cross-indication effects of physician word of mouth in a very operational framework.

PATIENT FEEDBACK BETWEEN INDICATIONS

As the number of patients diagnosed with a particular disease increases, the increased frequency of interaction with the health care system may result in other conditions being identified. As patients experience more *visits to the physician*, for example, it is possible that other comorbid conditions are recognized and diagnosed, as is shown in Figure 7.

The diagnosis feedback effect shown in Figure 7 can be used to develop a positioning strategy for an individual product. If certain comorbidities are common in a particular indication, a marketing campaign may help physicians identify such conditions. For example, physicians could be encouraged to look for specific symptoms in particular groups of patients, indicating previously undiagnosed comorbidities. The positioning strategy of a particular product may attempt to leverage such dynamics by educating physicians on potential overlap between indications.

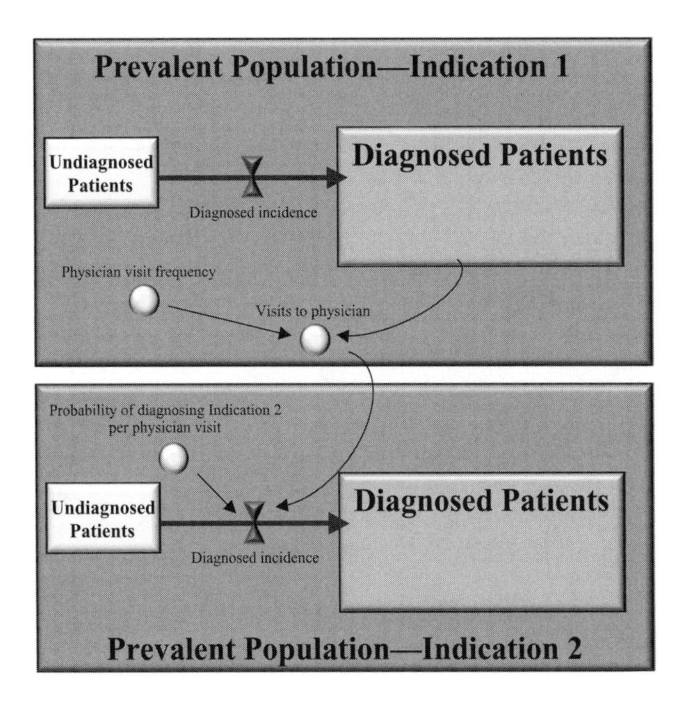

Figure 7 Example dynamic model structure of patient feedback between indications.

By explicitly accounting for overlap and interaction between indications, a portfolio model that incorporates Standard Template models of various diseases provides a unique and valuable approach to analyzing multiple markets. By incorporating dynamic interrelationships, such portfolio models provide a level of insight and rigor not commonly seen in simple roll-ups of traditional static forecasting efforts.

Portfolio analysis across indications involves combining and adapting Standard Template dynamic models to evaluate a firm's overall strategic position. This type of portfolio examination utilizing the Dynamic Modeling approach can be extended backwards to investigate the drug discovery and development process that ultimately brings potential compounds to the marketplace. Chapters 16 will address the methodology of this extension in greater detail.

SUMMARY

- Master or enterprise models that combine assessments of individual disease marketplaces are an important part of portfolio management and strategic development.
- Disease markets rarely exist in isolation, so integrating them requires some finesse in the application of methodologies.
- Comorbidity issues can overstate market potential by double-counting patients afflicted with multiple illnesses.
- Misdiagnosis by definition affects two indications, and the dynamics between correct diagnosis and misdiagnosis physically move patients between epidemiology categories.
- Feedback mechanisms affecting patients and/or physicians can impact how cross-indication dynamics play out over time.
- The Dynamic Modeling framework can be applied to create enterprise models that explicitly account for the interactions of effects between disease markets, resulting in an overall portfolio model that encompasses all relevant indications in a company's strategic plans.

16

Dynamic Modeling of Pharmaceutical Pipeline Portfolio Issues*

INTRODUCTION

This chapter will cover the following topics:

- Overview of the portfolio problem
- How Dynamic Modeling approach can be applied to pipeline portfolio issues
- Continuous versus agent-based models
- Analysis of pipeline dynamics
- Advantages of Dynamic Modeling in pipeline portfolio analysis
- Extensions of basic stock/flow compound pipeline framework
  - Move to OTC
  - In-licensing
  - Out-licensing
  - Line extensions (LEs)
  - Incorporating parameter uncertainty
  - Revenue streams
  - Resource requirements

Portfolio management concepts have been an accepted component of the financial services realm for decades, and the need to diversify risk continues to be an effective organizing principle for many financial instruments and analytic techniques. Risk is also a key component in the pharmaceutical world, where as few as 1 in 5000 laboratory-tested compounds ever makes it to the market (1). As such, the pipeline portfolio is a critical indicator of the future of a pharmaceutical

* Our special thanks to our colleague Kirk Solo, whose pioneering work in the application of Dynamic Modeling to pipeline portfolio issues is reflected throughout this chapter and in the book *The Process of New Drug Discovery and Development 2nd Edition* (Smith, C and O'Donnell, J. Informa Healthcare, 2006).

firm, and the process of analyzing and managing it effectively is of vital strategic importance.

Pharmaceutical pipeline portfolios, however, have industry-specific nuances that present some significant challenges. Traditional financial portfolio analysis often fails to address these subtle distinctions, while pharmaceutical domain experts often lack the financial portfolio background required to adapt those approaches to pipeline issues. The mismatch results in decision analysis based on a suboptimal set of evaluation tools. This type of inadequate analytic methodology (see chap. 1) has huge financial implications in the pharmaceutical world, where yearly research and development costs are now more than $50 billion globally (2).

The Dynamic Modeling methodology can be applied to pipeline portfolio issues in a way that integrates accepted financial portfolio concepts with the nuances specific to the development of pharmaceutical compounds. Such models serve as a litmus test for the future health of the enterprise by providing a useful and accessible tool for analyzing the possible evolutions of the product pipeline and understanding the implications of various strategic options for managing it. Many of our clients use dynamic models in this realm due to their ease of implementation, their sophistication of analytic techniques, and their flexibility to adapt to industry- or company-specific distinctions in the pharmaceutical industry.

ESTABLISHING A DYNAMIC MODEL FRAMEWORK FOR DRUG DEVELOPMENT

Previous chapters have shown how Standard Template models (or extensions thereof) can be applied at the level of indication-specific markets. However, entry into the marketplace is only one step of the drug development process, and many pharmaceutical companies struggle with the difficulties of pipeline evaluation and portfolio analysis. Fortunately, the Dynamic Modeling approach can be used to address these issues, and often begins with a stock/flow structure tracking compounds as they move through the development process, as shown in Figure 1.

This basic framework in Figure 1 shows the stages through which a new chemical entity (NCE) must go in order to reach the marketplace.* Any of these stages can be expanded to account for more detailed dynamics—the Discovery stage, for example, is often divided into subphases specific to an individual company's operating procedures. Figure 1 provides a visual, operational backbone of drug development that can be extended to meet to complexities of the actual pipeline process. The first and most obvious of these extensions involves failure of compounds in various stages of development. Adding these key components expands the diagram as shown in Figure 2.

* The approach described in this chapter can also be used for the evaluation of biologic pharmaceutical offerings in the form of new molecular entities (NMEs).

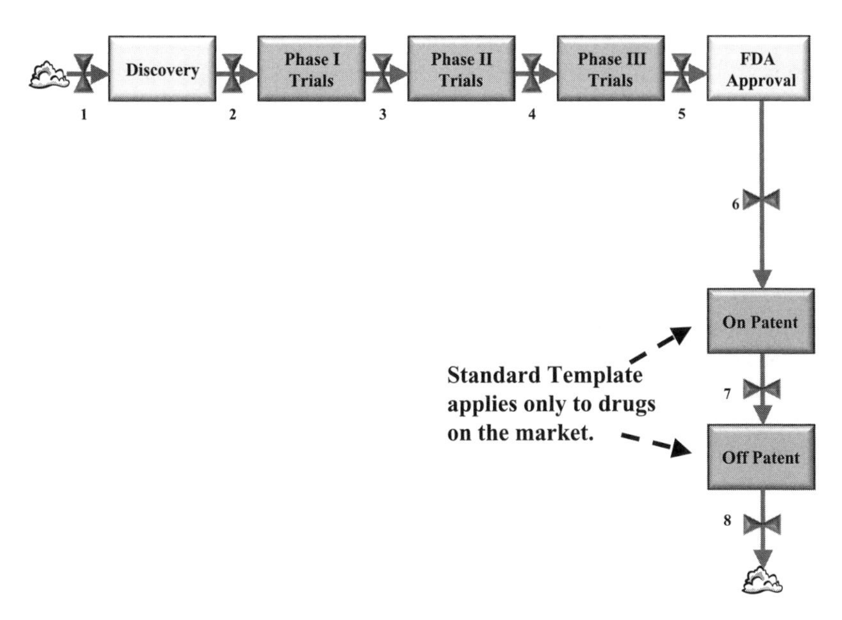

Figure 1 Basic stock/flow framework for drug development.

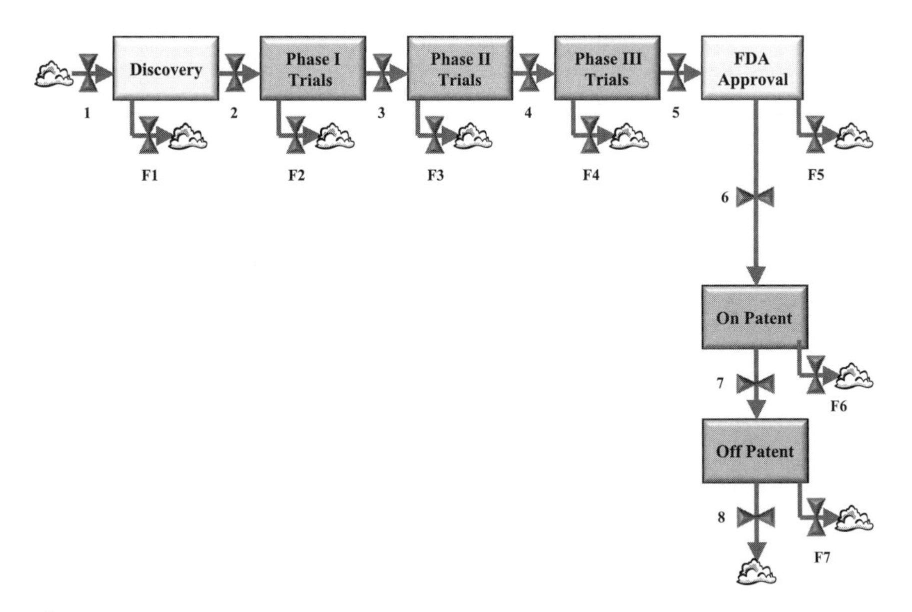

Figure 2 Drug development framework including compound failures.

The flows labeled *F1* to *F5* represent failures of drugs in various stages of development, including the approval process by the Federal Drug Administration (FDA). Flows *F6* and *F7* are intended to represent drugs being pulled from the marketplace due to adverse side effects instead of following the normal course of eventual "retirement" through flow 8.* Recall from chapter 3 that the clouds represent boundaries for the model, and a model with the stock/flow structure detailed in Figure 2 will no longer track any drug compounds entering into a cloud.Because of high compound failure rates, the drug development process is often thought of as a filtering process in which a large number of NCEs in the Discovery stage are needed in order to produce a viable, marketable compound in the On-Patent phase. The degree of filtering is large, with firms often managing hundreds of NCEs in order to produce a single revenue-generating market entry. For this reason, pharmaceutical firms often search for and develop multiple NCEs for a particular indication, knowing that most of the compounds will fail at some stage of their development, but hopefully at least one will ultimately be successful. Dynamic Modeling simulation can help evaluate the risk and uncertainty around such compound filter dynamics in the pharmaceutical pipeline to identify a range of possible outcomes for the pipeline portfolio.

By combining the concept of defined phases through which an NCE must pass with associated failure rates at those stages, Dynamic Modeling structure provides a consistent visual framework to depict a firm's drug development pipeline. The ability to simulate this structure makes the Dynamic Modeling approach to pipeline portfolio management even more powerful.

Some drug failures do not result in complete abandonment of the compound, and in certain cases NCEs may go back through various phases of testing to reevaluate their potential.† Problems with dosing, efficacy, and even side effects may dictate a retesting procedure (flows *R1–R3*), and including such dynamics into the stock/flow framework gives a more complete picture of potential developmental pathways, as shown in Figure 3.

As Figure 3 depicts, "failure" of an NCE is not necessarily a dead-end street. The possibility of retesting suggests some human element contributing to phase-test failure—the trial was not designed properly, the metrics proved to be inconclusive, and/or the dosing regimen was insufficient to show efficacy. Such human errors introduce the inefficiencies of rework into the drug development process, increasing accumulated costs and associated development time. Data-driven dynamic model simulation can help identify the impact of the retesting procedures that represent a strategic opportunity to improve the pipeline simply by reducing introduced errors into the filtering process.

* Readers may recall that in 2004 Merck pulled its painkiller Vioxx® from the market after a large clinical trial raised concerns about its long-term safety (3).

† The reevaluation process may identify new indications for the compound, based on analysis of results from failed tests. Eli Lilly's drug Strattera, for example, is now used to treat attention deficit/hyperactive disorder (ADHD) but failed clinical trials as a treatment for depression.

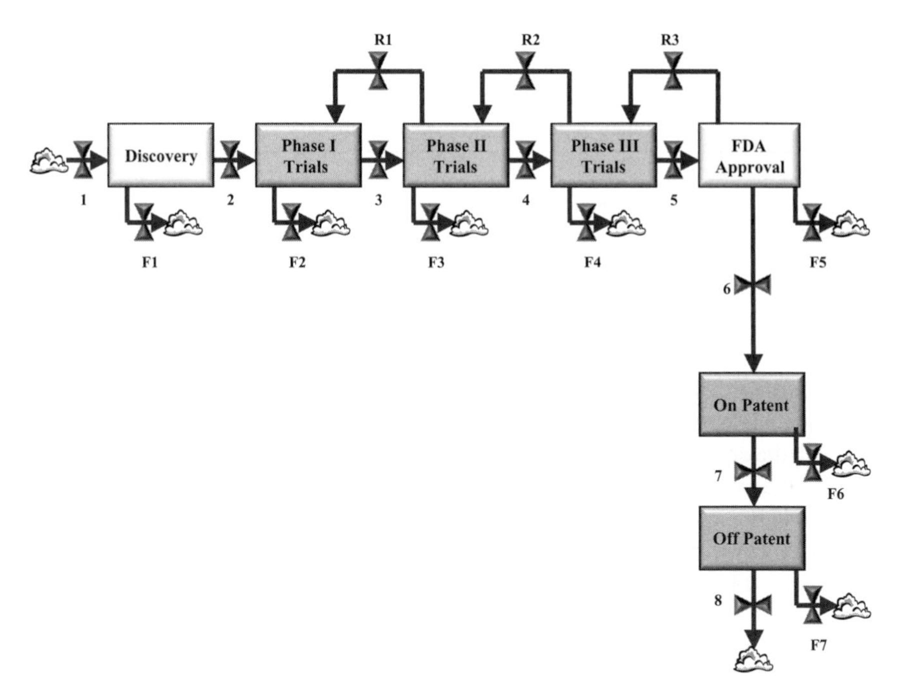

Figure 3 Drug development framework including retesting of phase trials.

POPULATING THE STOCK/FLOW FRAMEWORK

Pharmaceutical companies usually have multiple compounds in various stages of the pipeline with specific metrics around expected stage completion. The staging of compounds within the drug development pipeline is important for the following reasons:

1. As noted earlier in this chapter, the filtering aspect requires more NCEs in early stages then can be expected to ultimately enter the market.
2. Timing of compounds in various developmental phases affects resource requirements.
3. Eventual market entry and forecasted revenue generation impacts a firm's financial outlook.

For example, a pharmaceutical firm may have information on the status of its drug development process as shown in Table 1.

Table 1 shows aggregate metrics for the time spent per stage, although in reality most firms have expected completion dates specific to individual NCEs. Table 2 shows an example of how NCEs in the Phase III process might be categorized in terms of their expected completion dates.

To this point, our modeling examples have not included defined time metrics such as those in Table 2, but advances in Dynamic Modeling simulation technology

Table 1 Example Data Regarding Pipeline Status and Associated Time Metrics

Stage	Compounds in stage	Average time in stage (yr)
Discovery	20	3
Phase I	12	2
Phase II	10	1.5
Phase III	4	2
FDA approval	2	1.5

make this type of data incorporation relatively easy. Recall from chapter 9 that a dynamic model can uniquely account for and track individual NCEs as they move through the development process, including compound-specific parameters, which govern their eventual behavior. The ability to handle relatively small numbers of items (NCEs, in this case) discretely within the Dynamic Modeling construct is a powerful way to integrate the overall approach with the details required to analyze drug development portfolios effectively.

MODELING TERMINOLOGY: CONTINUOUS VS. AGENT-BASED MODELS

Most of the examples from previous chapters showed how the Dynamic Modeling approach can be applied to large groups of patients and/or doctors as they exist within a dynamic system. Such models are categorized as **continuous**; they count items in aggregate and do not differentiate between individual elements. From a technical standpoint, continuous models place items in stocks and subsequently move a certain number or fraction of them through the associated flow dynamics into other stocks. This approach requires that the items be grouped together; individual elements are not distinguished or tracked. Continuous models can be disaggregated into Segments or Arrays (see chap. 5), but segmentation to the level of unique elements with individual characteristics is generally too detailed to be useful in the realm of overall market dynamics as captured in the extended Standard Template. Chapter 9 introduced a technique called agent-based modeling

Table 2 Example Product-Specific Completion Date Data for Phase III Compounds

NCE	Expected phase III completion date
Phase III.1	March 2011
Phase III.2	December 2011
Phase III.3	June 2012
Phase III.4	August 2015

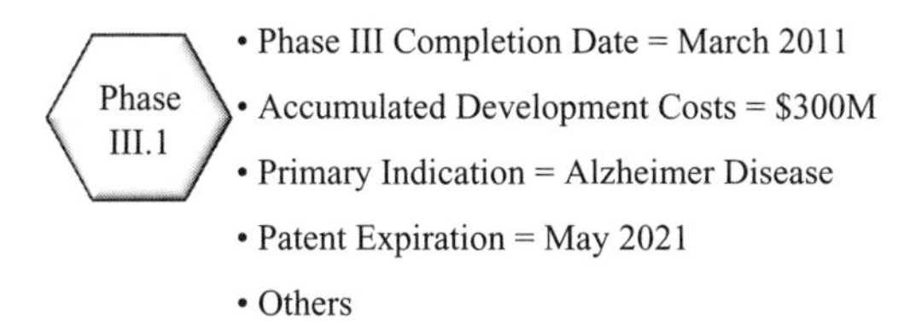

Figure 4 Conceptual model agent with associated tracking metrics.

(ABM) and gave some examples of how such an approach is used in the realm of patient and physician dynamics. In a similar fashion, the ABM methodology provides the degree of item-specific detail necessary to effectively analyze drug pipeline issues.

Most pharmaceutical firms have NCEs pipelines totaling less than 1000 and often no more than 40. Handling these smaller sample sizes requires the ABM approach that is capable of attaching individual characteristics to unique, discrete elements. Recall that in an ABM, a simulation environment is created in which individual agents represent actual NCEs in the pipeline, complete with unique characteristics that can be compound specific.

A simulation agent in a pharmaceutical research and development process can be thought of as a stand-alone entity with individual parameters that govern or track its behavior through the pipeline, as conceptualized in Figure 4.

The stock/flow structures in Figures 1 to 3 can be populated not with generic numbers of compounds and aggregate metrics, but rather with agents representing actual NCEs in the pipeline and their associated individual characteristics. A portion of the overall stock/flow chain populated with sample agents is shown in Figure 5.

Establishing an ABM simulation gives the Dynamic Modeling approach the flexibility and specificity necessary to effectively analyze pipeline dynamics consisting of specific, unique NCEs. For diagram simplicity, the icons for individual

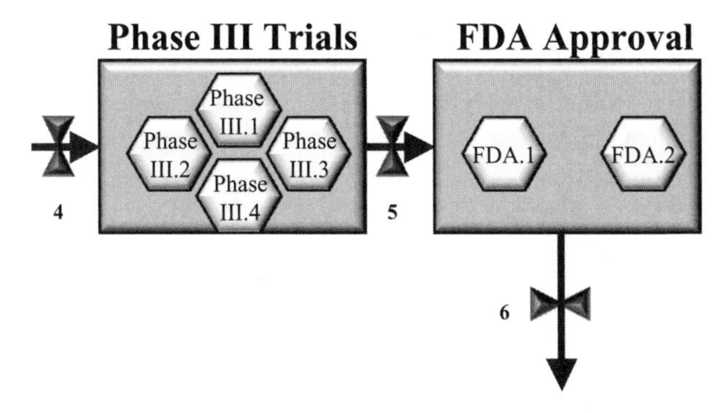

Figure 5 Portion of drug development framework with sample agents.

Table 3 Example Data for Stage-Specific Compound Failure Rates

Stage	Associated flow	Failure rates (%)
Discovery	F1	60
Phase I	F2	30
Phase II	F3	20
Phase II	R1	20
Phase III	F4	25
Phase III	R2	10
FDA approval	F5	10
FDA approval	R3	15
On-patent	F6	2
Off-patent	F7	0

agents will no longer be shown as this example is extended, but they remain the basis for both the modeling approach and all subsequent analysis.

INCORPORATING COMPOUND FAILURE RATES

One of the difficulties in pharmaceutical pipeline management lies in quantifying the impact of failure metrics of NCEs in various stages of development. Without some sort of analytical tool to incorporate such risks, the behavior of the overall pipeline is nearly impossible to intuit. Collecting either internal estimates or industry statistics on compound failure rates at different points in the drug development cycle and incorporating them into the Dynamic Modeling framework can help managers better understand the possible behavior of their current and future pipeline. Data from the Tufts Center for the Study of Drug Development (http://csdd.tufts.edu) is the industry standard for initial estimates regarding compound failure rates, and often provides a base from which a pharmaceutical firm can make an informed judgment regarding the metrics governing the behavior of their individual drug pipeline. For example, a team in charge of pipeline analysis may estimate a set of stage-specific failure probabilities as shown in Table 3.*

The estimates in Table 3 indicate 60% of drugs in the discovery phase never make it to Phase I (defining the flow of *F1*). Also, 25% of compounds never make it to the FDA-Approval phase from Phase III (*F4*), but an additional 10% have to go backwards in the process to Phase II trials (*R2*). These data can be incorporated directly into the stock/flow diagram to have a complete picture of the

* This data is presented in aggregate form, as if these parameters applied to all NCEs in the pipeline. Compound failure rates can be NCE-specific, however, and tied to individual model agents representing a particular compound.

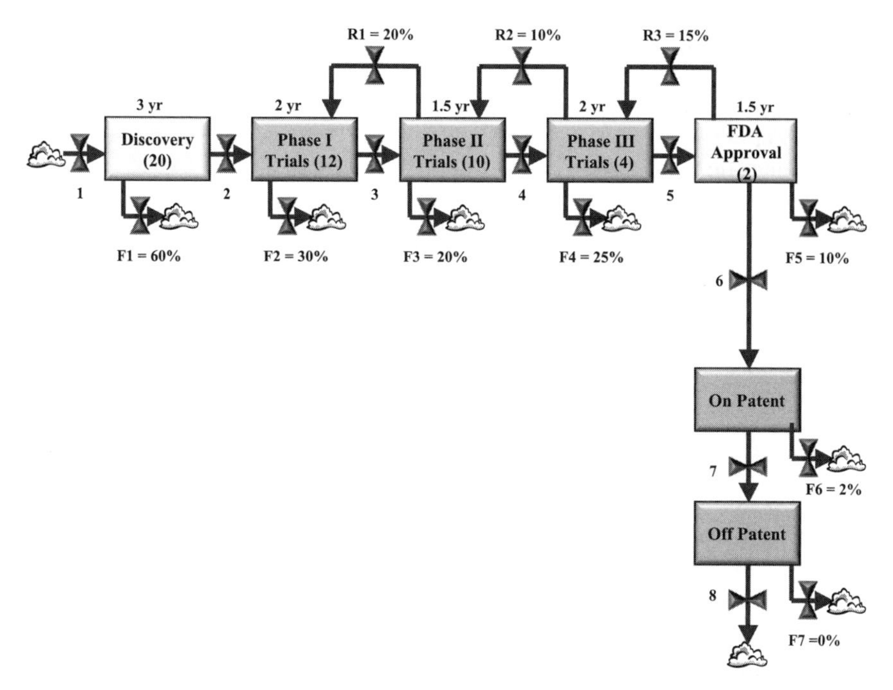

Figure 6 Drug development framework with associated data parameters.

structure and associated parameters for the drug development pipeline, as shown in Figure 6.*

ANALYSIS OF PIPELINE DYNAMICS

Even with fairly simple structure and associated data shown in Figure 6, the series of static calculations needed to analyze the possible behaviors of the drug development pipeline over time are overwhelming. However, Dynamic Modeling software has the capacity to simulate such a model and determine a range of possible outcomes, given the set of input parameters governing the behavior of the system. The most common and insightful simulation result in these types of discrete models is **Monte Carlo analysis**. In a Monte Carlo experiment, a large number of simulations are performed independently and the corresponding results are grouped together for analysis. In effect, each simulation outcome is unique as it represents the results of a specific set of input assumptions. But these input assumptions are drawn from a distribution attached to individual agents (representing actual NCEs) for each simulation, as in the case of compound failure rates. For example, in 300 of 1000 simulations would a particular NCE fail the

* *Remember*: Some of the parameters governing the NCEs in the pipeline reside at the agent level and are not depicted in this type of aggregate diagram.

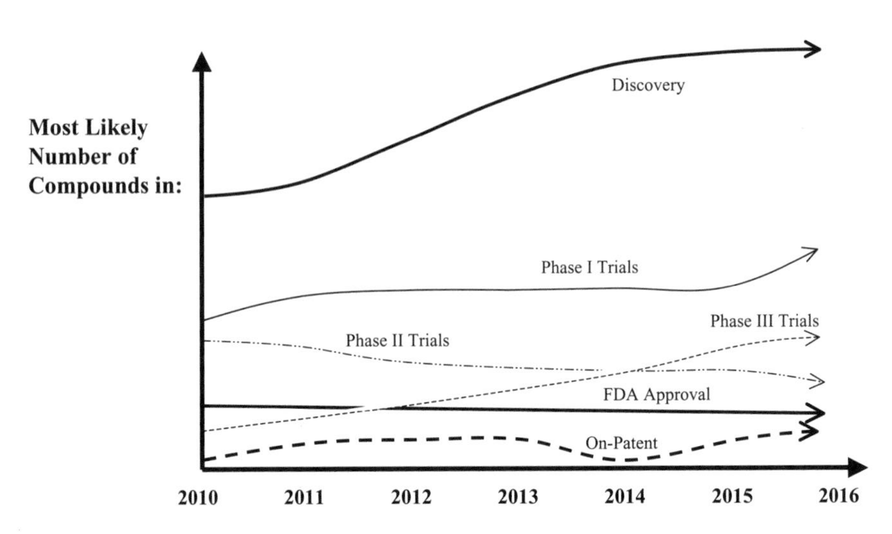

Figure 7 Example most likely trajectory of compound stages.

Phase I Trials stage (30%— Table 3). The individual agents of the dynamic model are parameterized as the beginning of each run of a multiple simulation Monte Carlo analysis, producing a range of possible outcomes. In this manner, Monte Carlo analysis provides an analysis of the overall variability of model results based on a structure and a defined set of input assumptions attached to individual agents.

To continue with this example, assume that the flow of compounds into the Discovery stage (flow *1*) is 10 NCEs per year. In other words, the firm's scientists identify 10 new compounds, which will enter the Discovery process (which lasts, on average, for 3 years) each year. Given this input to the system and the overall times associated with stage completion and associated failure rates, what is the range of outcomes of the firm's drug pipeline over time? This example will assume that no drugs are currently in the On-Patent or Off-Patent stages, although including such on-market compounds is an easy extension.

Figure 7 shows the most likely outcomes for the firm's pipeline, based on a Monte Carlo analysis of the dynamic model established in this chapter.

Charts such as the one in Figure 7 are useful in analyzing the likely evolution of a pharmaceutical firm's product portfolio over time, as they demonstrate the most likely number of compounds in any development stage over any given interval. In addition, the analysis can be extended to show the distribution of outcomes along these same dimensions. These patterns give aggregate evidence as to likely evolution of the pipeline, and often suggest associated resource requirements and medium-range financial planning issues.

As noted earlier in this chapter, however, "most likely" does not imply certainty, and a more illuminating result from a Monte Carlo analysis is the range surrounding an expected result. For example, the projection of the number of

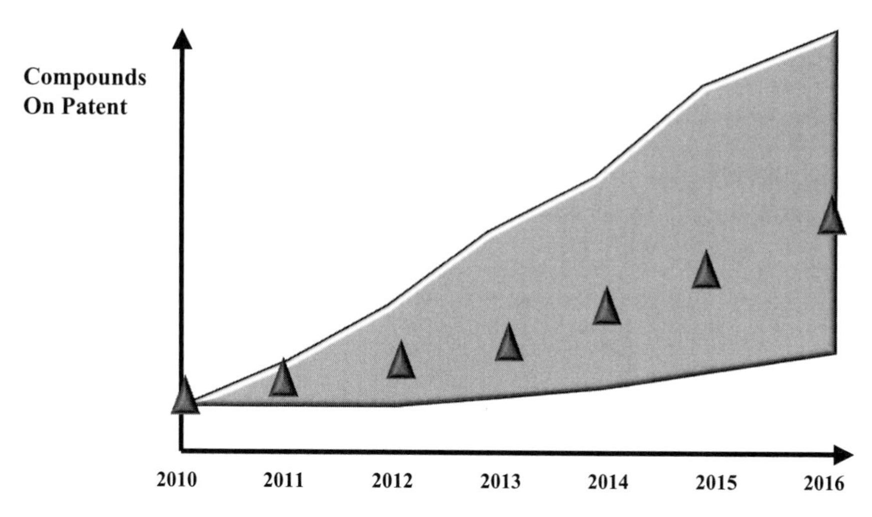

Figure 8 Example revenue trajectory uncertainty.

compounds On Patent can be shown with a distribution around expected outcomes as shown in Figure 8.

Although the results for the On-Patent stage are shown in Figure 8, the dynamic model can capture the resulting data from any of the defined phases in the stock/flow diagram. The results of this type of Monte Carlo simulation can be analyzed to produce a series of useful metrics, such as:

1. *Confidence intervals*: What percentage of results fall above or below a certain threshold?
2. *Standard deviation*: How wide is the distribution of results around the expected or most common outcome?
3. *Causes of variability*: Which parameters, such as average time spent in each stage or compound failure probabilities at various phases, contribute the most to the distribution of overall outcomes?
4. *Sequencing of adverse outcomes*: Which combination of factors contributes to the outcomes that differ the most from the expected result? Because a dynamic model saves the particular set of inputs that generate a specific output, any outliers in the distribution of results can be "deconstructed" to discover the series of events leading to that interesting result. This sort of model post-mortem often identifies areas that warrant further investigation.

A careful examination of these types of analyses leverages the Dynamic Modeling simulation methodology with a Monte Carlo approach and can help portfolio managers better understand how their pipelines might evolve over time. This more complete understanding can then be translated into better strategic decisions regarding portfolio management and the strategic decisions surrounding a firm's developmental pipeline.

DYNAMIC MODELING VS. ALTERNATIVE ANALYSIS TOOLS

Some readers may have seen pharmaceutical portfolio analysis designed and implemented in a spreadsheet or other computational applications, including the ability to compute Monte Carlo results. While these approaches can indeed be of value, our clients have seen a number of additional insights into pipeline dynamics that *only* a Dynamic Modeling simulation can provide.

Advantages of Dynamic Modeling in Pipeline Portfolio Analysis

- Dynamic models are more flexible in their establishment of structure/metrics and are not restricted to the financial metaphor underlying most other approaches.
- Stock/flow methodology provides a visual framework to conceptualize the pipeline development process.
- Operational diagram can be populated with simulation agents representing individual compounds in various developmental stages.
- Agents are easily aggregated, extended, and duplicated, making dynamic models easy to appropriately scale and subsequently analyze.
- Compound-specific attributes can be assigned to agents, which may then be inherited by LEs generated from parent compounds.
- Interaction between NCEs is easily captured, as in cases where success/failure of one compound influences the expected success/failure of another.
- Dynamic model structure can be simulated to determine the expected outcomes from a set of associated input assumptions, both in terms of existing pipeline and future portfolio scenarios.
- Condition of the pipeline can be observed at any time point by simply pausing the simulation and observing the status of the system.
- Dynamic models explicitly categorize the causes for model results instead of simply calculating combinations of statistical distributions. Analysis of simulation results can categorize not only *what* happened but *why* it happened as well.
- Stock/flow structure combines uncertainty with cause-and-effect relationships, which can be quantified through simulation.
- Dynamic simulations can categorize system variability, including the effect of strategic decisions on changes in metrics over time.
- Methodology can incorporate market models of revenue generation for individual compounds versus simple scenario assumptions.
- Supply/production issues can be integrated with pipeline scenario evaluation.

Many of these advantages will be demonstrated throughout this chapter as the basic Dynamic Modeling framework for pipeline portfolio analysis is extended.

Dynamic Modeling of Pharmaceutical R&D Processes

Sue E. Steven, PhD, MBA
Global Programs and Portfolio Management
Invitrogen, Inc.
Former Head of Portfolio and Resource Management, Centocor, Inc.

Although the authors have clearly made their case for use of Dynamic Modeling in the pharmaceutical R&D environment, Centocor, Inc., had some additional requirements that this type of modeling best supported. These include

1. a modest portfolio,
2. an exponentially growing portfolio,
3. a predominantly biologic portfolio,
4. the need to integrate existing portfolio with future compound generation portfolio, and
5. the need to integrate with larger Johnson & Johnson portfolio.

Understanding the output of this kind of modeling helped us to set strategic targets and ensure our portfolio decisions were in alignment with our targets. The following text provides more details:

1. A modest portfolio benefits from Dynamic Modeling as it can best represent the vagaries of a single event skewing the outcome of a portfolio. The expected NPV of a single asset is often portrayed by simply risk adjusting the NPV (e.g., 50% probability of success multiplied by a $200M NPV = $100M expected NPV). Yet we know the outcome for the asset is either $0 or $200M, not $100M. Just so, a small portfolio can be subject to the same inappropriate approximation for both expected revenue and expected investment unless you use Dynamic Modeling.
2. All pharmaceutical companies expect to grow their pipelines in order to maintain double-digit revenue growth, but only small companies can expect to grow their pipelines exponentially in order to become a major player in the business. With that rate of growth, and looking at a 10-year time period, the percent of the pipeline that is "yet to be discovered" is significantly larger than the existing pipeline of projects in R&D. Thus, there is a greater dependency on understanding year-by-year estimates on the requirements needed to grow such a portfolio. These include the financial investment and FTE resources required at both at the individual department as well as the company level.
3. Having a predominantly biologic portfolio poses several issues that Dynamic Modeling can address.
   a. The biologic therapeutic industry is young, there are inadequate industry benchmarks, and even less within company/executive experience to fully understand the risk issues and the uncertainties of compound

development time, required investment, and subsequent revenue generation. Dynamic Modeling is an easy way to try out different assumptions regarding these crude benchmarks and understand the ones that impact the portfolio the most.

b. With a growing biologic pipeline, pharmaceutical development, technical transfer, scale up, and supply chain capacity often becomes the rate-limiting factors in achieving revenue growth. Dynamic Modeling can be used to determine the degree of uncertainty around the demand for product for which capacity is being built. Since the decisions to build/buy plant capacity must be made as early as five years before the product launches in the marketplace, early insight into capacity and supply chain decisions is extremely valuable.

4. All established companies want to see the integration of their existing R&D portfolio (projects already underway) with the future projects that have yet to be "invented/imagined". Dynamic Modeling can provide the representation of both portfolios together, as well as the individual contribution of each. The existing portfolio has committed resources that are easily identified, but the input assumptions driving the future portfolio can be altered to test their impact on the integrated portfolio and associated required resources over time.
5. Finally, Centocor, Inc., needed to integrate its biologic portfolio with the larger J&J pharmaceutical portfolio that was also modeled dynamically but with assumptions more appropriate for small molecules. The total J&J portfolio could then be examined in several different ways. Scenarios could be run to see the impact of an increase in investment to the entire integrated development process or the effect of reallocating the R&D budget from one J&J Company to another. Alternatively, the implications of a reduced cycle time for a particular development phase could be analyzed and evaluated. Such analyses provided key insights into how company resources could best be deployed.

In conclusion, Dynamic Modeling is an effective, low-cost method of examining different scenarios to analyze the future outcome of a pharmaceutical portfolio. This process enables the strategic targets set by management to be defined more clearly and supported by effective resource allocation.

EXTENSIONS OF THE BASIC STRUCTURE

The basic stock/flow structure and the associated simulation model of pharmaceutical pipeline management provides a useful framework to analyze basic drug development systems. Readers familiar with these issues, however, know that the scope of actual pipeline portfolio management is much more complex. Fortunately, the operational simulation framework established in this chapter is flexible and well suited for some key additions. Specifically, seven separate extensions have proven themselves valuable to our clients. These extensions will be shown in here in isolation, but can easily be integrated to develop a comprehensive dynamic model of the entire portfolio management process.

These seven common extensions of the basic stock/flow pipeline structure are listed below and discussed individually in the text that follows:

1. Taking a compound to OTC market
2. In-licensing compounds from other firms
3. Out-licensing compounds to other firms
4. Including LEs from parent compounds
5. Incorporating probability distributions for input parameters
6. Calculating associated revenue streams using the Standard Template
7. Computing resource requirements

Taking a Compound to the OTC Market

Firms with compounds in the Off-Patent stage (or heading that way) are increasingly turning to the over-the-counter (OTC) market as a potential strategic option for compound lifecycle management. Advil is perhaps the most successful of these OTC transitions, but in recent years brands such as Claritin, Nicorette, and Benadryl have made similar conversions. The details of evaluating, managing, and implementing are myriad, but a simple extension of the basic stock/flow framework provides a means to test and analyze the revenue effects of this strategy. Figure 9 shows how the dynamics of the move to OTC can be captured in stock/flow methodology.

The auxiliary variable, i.e., *OTC revenues per compound* is of course not a simple constant metric, and a complementary analysis is often done to determine the scope and magnitude of the factors which affect it. Some of these influencers might be

1. size of the associated OTC market;
2. number and aggressiveness of existing competitors;
3. effectiveness of the compound in the indication;
4. marketing spending in support of the compound while on the OTC market.

A comprehensive analysis of these factors is beyond the scope of this work, but these influencers are often incorporated into the dynamic model to more fully address the revenue implications of a move to the OTC market for a given compound.

In-Licensing Compounds

Pharmaceutical firms often choose to in-license compounds to beef up their developmental pipeline or round out their existing product portfolio. Structurally, in-licensing can be represented with a simple extension of the basic stock/flow framework, as shown in Figure 10.

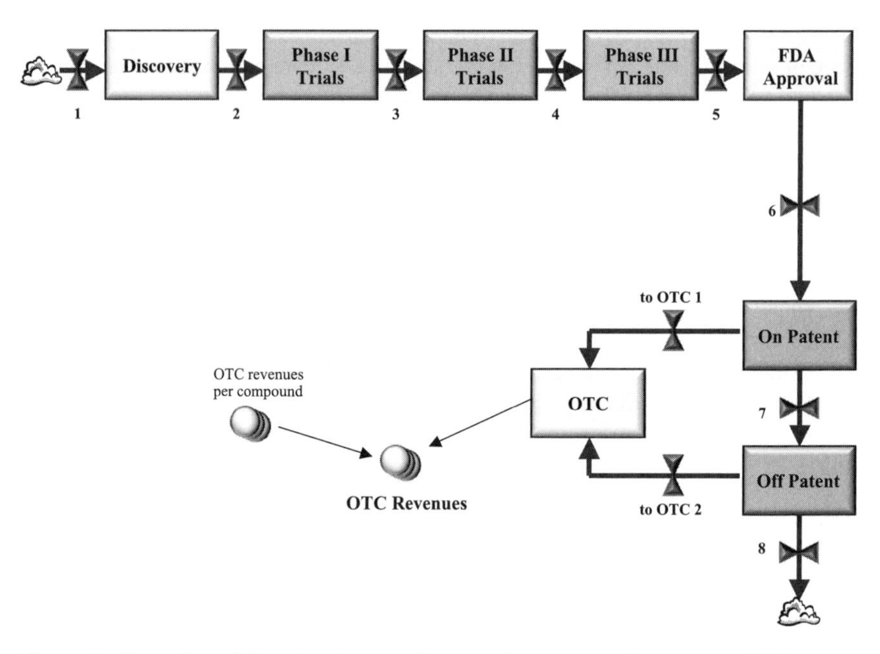

Figure 9 Extension of drug development framework to account for move to OTC.

In-licensing opportunities can enter the drug development process at the level of Phase Trials and/or the On-Patent stage. In either case, the simulation model can assign a cost necessary to obtain in-licensed compounds (*compound acquisition cost*) that can be compared to their expected revenue streams (determined from Standard Template dynamic models) to determine the financial return of these types of in-licensing strategies. Once entering the stock/flow model, in-licensed compounds are assigned dynamic metrics, which govern their subsequent behavior. These parameters can be the same as those for internally generated NCEs, or in-licensed compounds may be given differential probabilities and tracked separately as they advance through the development pipeline. Some of our clients, for example, assign higher failure probabilities for in-licensed compounds, reflecting a higher inherent level of uncertainty regarding their changes of success. Monte Carlo analysis, as described previously in this chapter, is then performed to determine the range of possible outcomes for this set of in-licensed opportunities.

The timing of in-licensing is an important aspect in evaluating potential external opportunities. In a world of diminishing pipelines, the in-licensing of late-stage compounds has become an expensive game to play. As pharmaceutical firms are forced to evaluate in-licensing opportunities further and further back in the development pipeline, the need for careful analytics expands. One nice feature of the Dynamic Modeling approach is the ability to bring external NCEs into the simulated pipeline structure at any desired stage of development and then subject those agents to specific parameters governing their subsequent behavior.

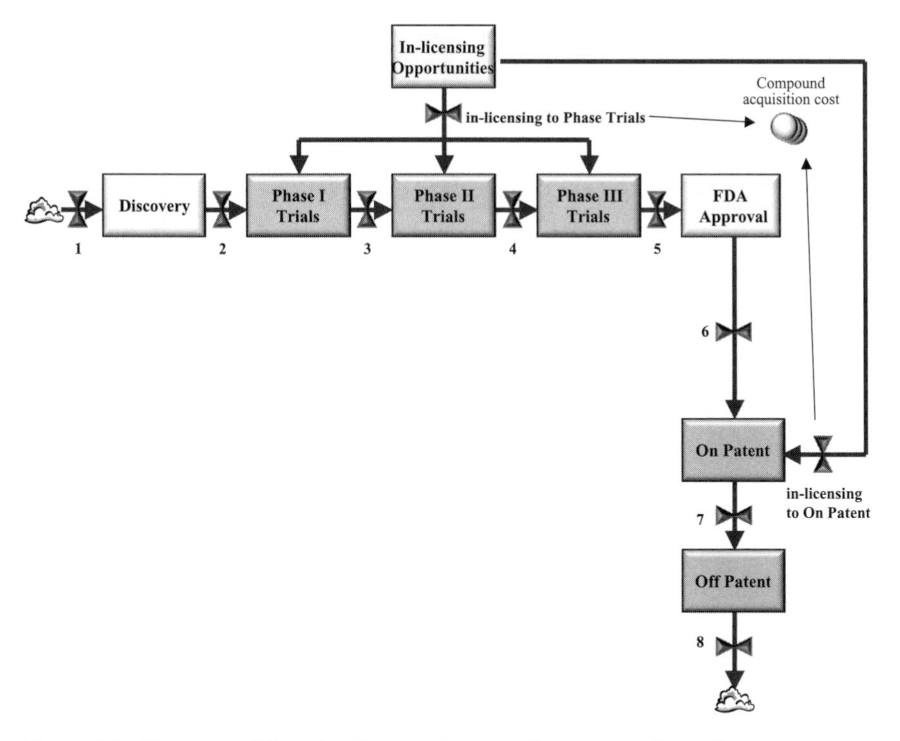

Figure 10 Extension of drug development framework to account for in-licensing.

Out-Licensing Compounds

Out-licensing opportunities represent another strategic option for firms to decrease the number of compounds in their pipeline and/or generate immediate cash by "exporting" compounds to other firms. The possible revenue generation from and the strategic nuances to out-licensing are varied, and can be analyzed with the help of an extension to the basic stock/flow framework as shown in Figure 11.

Typically, revenue streams from such divestitures come in two forms: (*i*) an acquisition fee paid by the acquiring company for the right to develop the acquired compound (*compound acquisition revenues*) and (*ii*) some sort of revenue sharing arrangement if and when the compound reaches the market (*out-licensed revenues*). The variable *revenues per out-licensed compound* can be computed using a Standard Template dynamic model, or more commonly, a time-based scenario of revenue generation over the life of the compound before it eventually exits through the *retirement flow*. A proper analysis of the *revenues per out-licensed compound* flow should incorporate the uncertainty of market entry, however, if the out-licensed NCE is one that has not yet passed the FDA-approval process.

Pharmaceutical firms are understandably hesitant to out-license their compounds, often due to lack of clarity regarding the strategic and financial

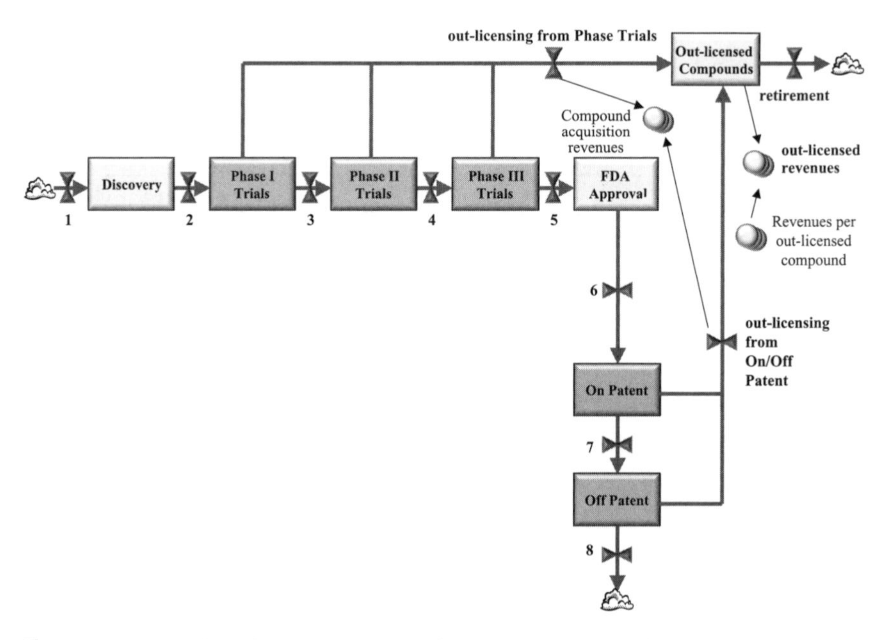

Figure 11 Extension of drug development framework to account for out-licensing.

implications of such actions. Fear of creating competition in an important indication, unwillingness to depart with viable compounds, and uncertainty around the ultimate impact of out-licensing decisions certainly make the case for caution on the part of portfolio managers. Including out-licensing possibilities into the dynamic simulation model, however, provides the means to analyze and quantify the potential pros and cons of such options in a consistent and clear manner. The uncertainty surrounding out-licensing will never disappear, but the analytical framework of a dynamic pipeline portfolio model can be used to comprehensively evaluate this strategic option.

LEs from Parent Compounds

During the development process or even while in the On-Patent stage, drug compounds sometimes develop what are called LEs: new indications, methods of administration, or even dosing regimens which create new business opportunities for existing compounds. LEs generally have to go through the same testing phases as NCEs, although they may not necessarily have to begin at Phase I. Because LEs are "offshoots" of parent compounds, they do not replace the NCEs from which they are spawned. Instead, LEs should be treated as newly generated compounds that retain some of the properties of their parent compound (patent expiration, accumulated costs, failure of the parent compound triggering failure of the LE, etc.) Incorporating these dynamics is fairly straightforward from a structural standpoint, as shown in Figure 12.

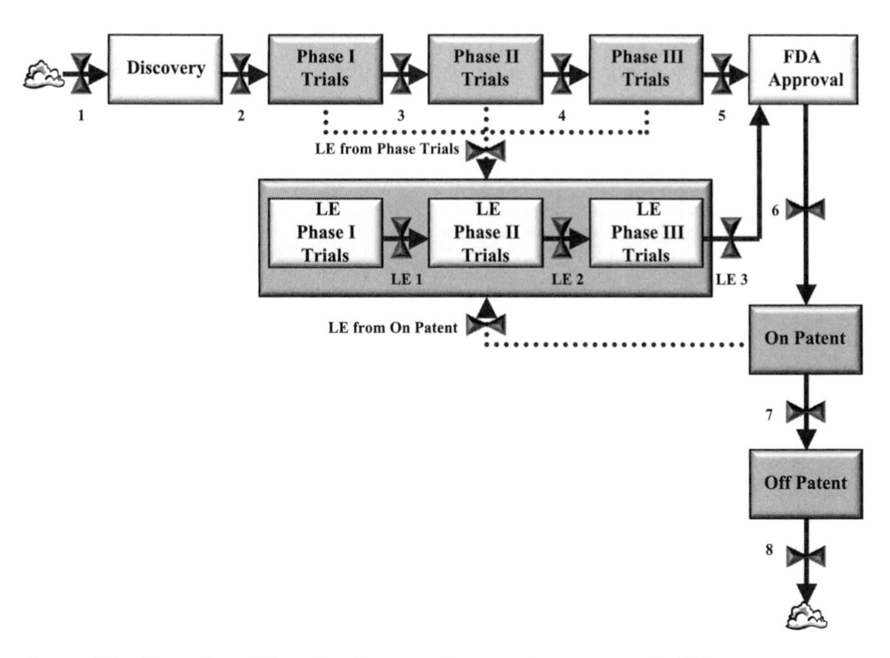

Figure 12 Extension of drug development framework to account for LEs.

The dotted lines on the flows of *LE from On-Patent* and *LE from phase trials* in Figure 12 denote that existing parent compounds are not actually physically transported back to LE phase trials when an LE is created. Instead, a new agent is created in the model that retains certain properties from the parent agent, and this new agent is then sent along the path of LE phase trials. The subsequent behavior of the LE will be governed by how this new simulation agent is parameterized. Simulation of the LE dynamics gives a sense for the timing and implication of their inclusion in the pipeline process in a way that static approaches have difficulty capturing. Particularly in a pharmaceutical world where discovery of viable new compounds appears to be slowing, accurately portraying the dynamics associated with LEs will be increasingly important.

To fully incorporate the dynamics of LEs, a number of assumptions must be agreed upon and implemented in the Dynamic Modeling structure. These LE parameters often include, but are not limited to

- How often does a parent compound generate an LE?
- What testing phase does an LE enter?
- What are the residence times and failure rates for an LE in Phase Trials?
- Can an LE generate extensions from itself?
- Does an LE fail if the parent compound fails any of its testing phases or the FDA-Approval stage?

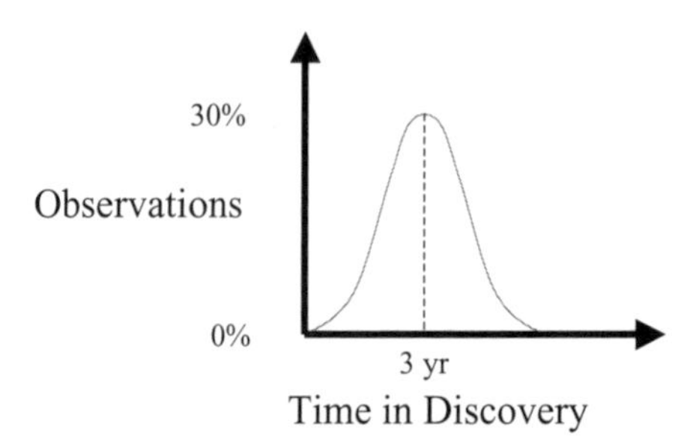

Figure 13 Establishing input parameters probability distributions (Example: Time in Discovery).

The inclusion of LE structure and the ability to quantify the parameters that govern its behavior provides a more realistic evaluation of drug development pipeline dynamics over time. The agent-based nature of Dynamic Modeling simulations makes this extension fairly easy to implement and allows for LEs to be analyzed as a distinct subset of the entire portfolio management process. The flexibility of this methodology allows for a very operational representation of the LE possibilities and seamlessly quantifies their inclusion in the overall drug pipeline.

Incorporating Probability Distributions for Input Parameters

As shown in Table 1 and Table 3, first-pass parameters for dynamic pipeline portfolio models are often assumed to be a single value, such as the average time in the Discovery process being three years. However, more advanced analysis of pipeline dynamics often includes a range around these mean values, introducing yet another layer of uncertainty and associated realism into the resulting dynamic simulations. For example, the parameter governing the time spent in the discovery phase can be established as a probability distribution, with a range of possibilities around a mean of three years, as depicted in Figure 13.

For each simulation and for each agent representing an NCE, the dynamic model will choose a value from the distribution to govern the time that particular NCE spends in the Discovery phase. Including this dimension of variability can often affect the overall distribution of potential outcomes as determined by Monte Carlo analysis and creates a simulation environment that more closely mimics the real world of drug development with uncertain governing parameters.

Readers with knowledge of statistics will note the importance that the left-hand "tail" of the distribution should not go below the "0" value of the X-axis. Doing so would allow the time spent in the Discovery phase to be negative—Clearly a troublesome input assumption. Conversely, the right-hand tail of many potential probability distributions can extend far beyond what is reasonable in the pharmaceutical world. For this reason, some client teams choose to establish minimum and maximum parameters for time spent in each stage, with some sort of

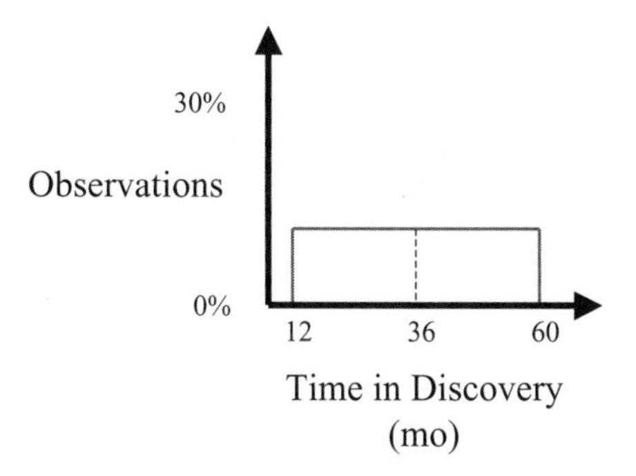

Figure 14 Establishing maximum and minimum for input parameter probability distributions (Example: Time in Discovery).

distribution curve between those points. Figure 14 shows an example of a uniform distribution between realistic minimum and maximum values.

The appropriate choice of distributions and their ultimate impact on overall portfolio dynamics is beyond the scope of this work, but Dynamic Modeling software is flexible enough to include a wide range of relevant probability distributions. Inclusion of uncertainty around input parameters is easy to incorporate into Dynamic Modeling simulations, and often adds a degree of robustness to overall model methodology and corresponding outputs.

The increasing availability of data has made the process of incorporating uncertainty, as an input parameter, much more robust and accessible for many portfolio managers. In our experience with clients, however, this data is more easily found than carefully incorporated. The rigorous framework of the Dynamic Modeling approach to pipeline issues and the associate simulation component for quantification is the best way to leverage the vast amounts of data related to these processes.

Calculating Compound Revenue Using the Standard Template

A primary goal of a pipeline evaluation is the impact on pharmaceutical revenues over time. Drug development is a risky world in which revenues from successful compounds must be used to offset and support the development costs for all the compounds that never make it to the winner's circle. Previous chapters demonstrated how Standard Template dynamic models, specific to individual drugs in particular indications, provide a solid framework for evaluating revenue potential upon entry into the marketplace. This approach works especially well for compounds currently in the drug development process, but can be extended to newly created compounds generated over the course of the simulation. Integrating the Dynamic Modeling results for particular indications with the pipeline simulation to determine probability of said compounds actually reaching the market results in an even clearer picture of the revenue possibilities in both the short and long term.

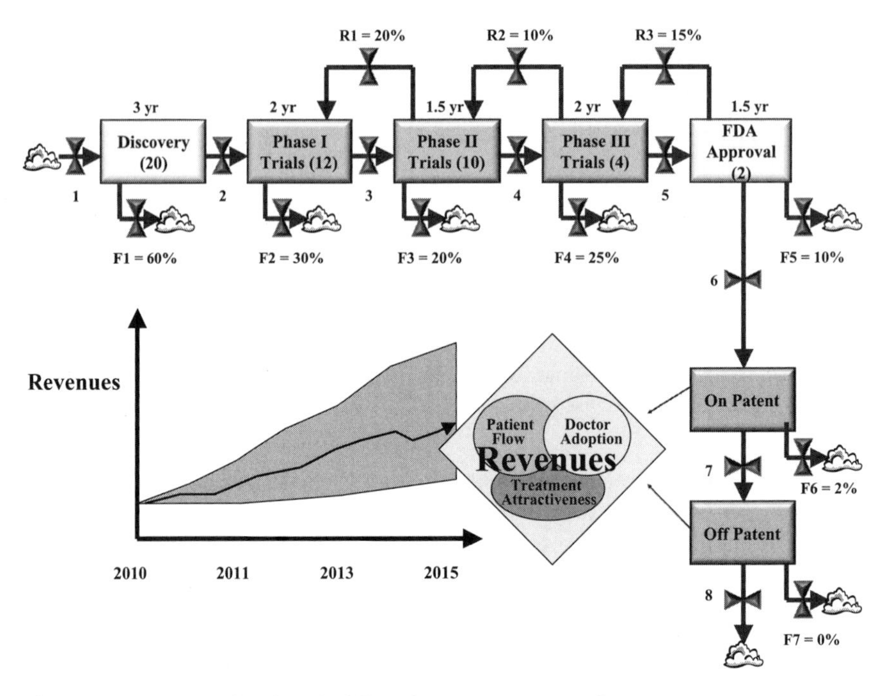

Figure 15 Incorporating Standard Template revenue generation.

Figure 15 shows how Standard Template models can be integrated with the drug development pipeline structure to determine a range of possible revenue curves over time.

In addition to integrating revenue calculations determined by Standard Template model, the dynamic pipeline simulation can accumulate the costs accrued to each individual compound as it moves through the development process. These costs, from both successful compounds and those that ultimately fail, can be compared to the projection of revenue generation to determine an expected profit over time. This analysis provides a range of possible financial metrics that are vital to long-term strategy development.

Computing Resource Requirements

The uncertainty contained in the pharmaceutical development process makes the associated resource planning task a daunting one. However, Dynamic Modeling simulations can shed some light on the resource requirement planning process by quantifying a range of possible outcomes from a firm's pipeline. These outcomes, usually measured in terms of number of compounds at various developmental stages over time, can be easily translated into the number of resources necessary over a certain time horizon. This staffing task is accomplished by identifying what are called **Resource Curves**—the number of full time equivalent (FTE) resources

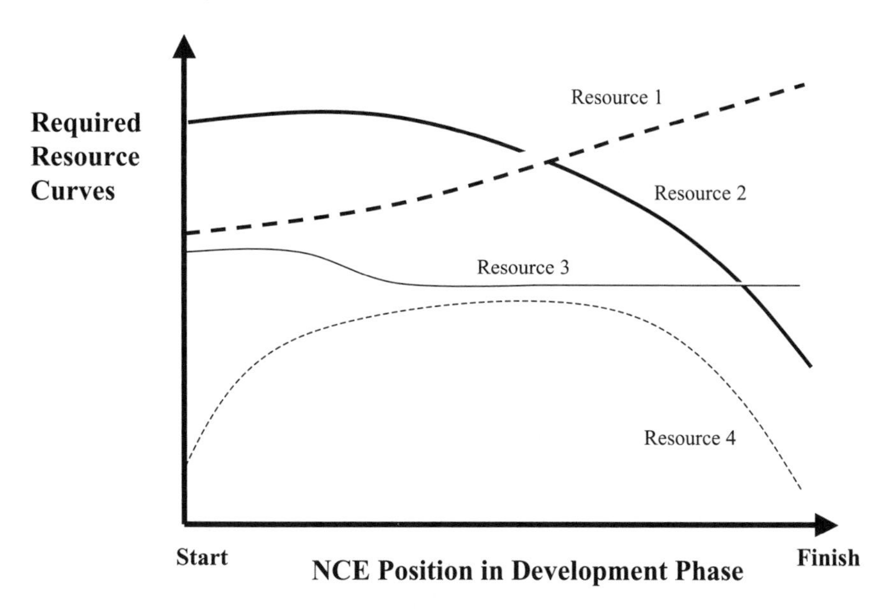

Figure 16 Example required resource curves for NCE development.

typically needed as an NCE advances through a particular development stage. For example, Phase II development may require a number of different specialized resources (up to 40 resource categories, depending on their groupings) in order to complete the necessary tasks. But those resources are employed at different rates throughout the NCE's advancement through that development phase. A dynamic model can incorporate this complexity relatively easily by first simulating an NCE as it moves through the Phase II development stage and then associating required FTEs (by resource category such as Program Manager, Biostatistician, Clinical Pharmacology Scientist, etc.) to that specific entity, as shown in Figure 16. Multiple simulation runs can then compute the range of possible resource requirements by each individual resource category, given the expected variability in the number of NCEs in any given development stage.

Dynamic Modeling simulation results of resource requirements provide a wealth of information for the purposes of human resources planning. Analysis focusing on various metrics of centrality and variation can be done at the development stage level, the specific resource category level, and/or at an aggregate firm level. Figure 17 shows an example of the expected number of FTEs required to meet a hypothetical pipeline scenario. Such resource requirement projections often identify important strategic gaps, which may require immediate attention. For example, a 25% increase in FTEs from 2012 to 2013 may not be possible, much less practical, given the current labor market. This analysis is often extended to include dynamics of up-to-speed time and productivity metrics associated with such rapid hiring. One possible strategy to manage this expected FTE gap to ramp

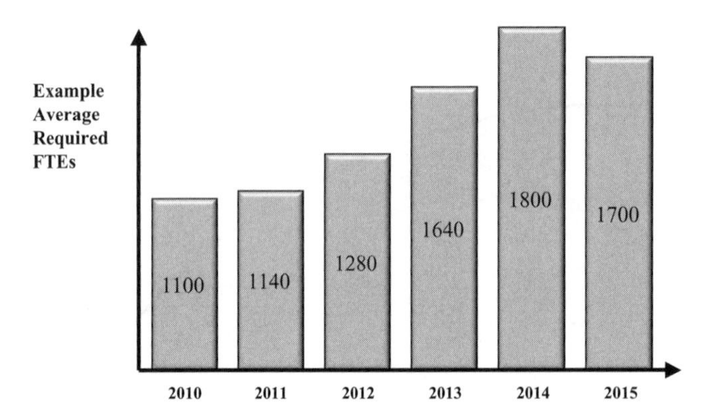

Figure 17 Example output regarding FTE requirements.

up hiring in the earlier years in anticipation of increased resource needs in the longer term. Alternatively, an out-licensing strategy of development compounds can be tested to reduce internal resource demands. Some firms use such projections to proactively pursue outsourcing partnerships with external resources being used as "swing capacity" in times of increased resource requirements. By providing an integrated means to determine a firm's resource needs over time, the dynamic portfolio simulation allows managers to systematically test and evaluate various strategic solutions and make informed decisions in a timely manner.

Strategy Tests Using the Dynamic Model Simulation

Dynamic models such as those described in this chapter can be used for a wide variety of scenario testing, what-if analysis, portfolio optimization, and strategy development. Our clients find these analyses both insightful and actionable, combining the power of well-established principles from the world of financial services with the power and flexibility of the Dynamic Modeling approach. The result is a framework that addresses the subtleties of pharmaceutical pipeline portfolio management—nuances that must be recognized and accounted for in effective strategic planning.

SUMMARY

- Dynamic Modeling methodology provides a unique way to address portfolio management problems in the pharmaceutical industry.
- The framework establishes a stock/flow structure representing stages that an NCE must pass in order to reach the marketplace.
- Associated failure metrics can be assigned to NCEs, either on an aggregate or individual compound level.
- Agent-based models create individual entities in the Dynamic Modeling framework to represent unique chemical or biologic compounds.

- Basic stock/flow structure related to drug development can be extended or modified to address various strategic options for NCE management.

REFERENCES

1. Pharmaceutical Industry Profile 2002, Pharmaceutical Research and Manufacturers of America (PhRMA)
2. On Treacherous Ground, 50 of the world's biggest pharmaceutical companies are challenged to perform in the most difficult environment ever. Med Ad News, September 1, 2002.
3. By Learning From Failures, Lilly Keeps Drug Pipeline Full. Wall Street Journal, April 21, 2004.

Glossary

Agent-based models: Dynamic models in which discrete simulation entities represent unique items, rather than collections or groups of elements. Often used for pipeline portfolio models in which agents represent individual drug compounds.

Arrays: A disaggregation option that can and often does exhibit interelement movement. Arrays allow for items (patients, doctors, compounds) to move or advance through stages defined as elements of the array.

Auxiliary: A building block of the Dynamic Modeling language where constants are stored, model inputs are housed, dependent relationships among variables are expressed, or calculations are made. Auxiliaries are not fundamental to system structure the way stocks, flows, and connectors are, but are useful in making dynamic models more explicit and testable.

Average residence time: The mean amount of time items spend in a stock before exiting to another part of a process or system, of particular interest when analyzing the relative magnitudes of stocks and flows in a system.

Balancing feedback: A feedback mechanism whereby a change in a model variable creates compensating reactions that come back to change the original variable in the opposite direction. Balancing Feedback are the control mechanisms that prevent systems from moving too far in one direction.

Bounding: A process of establishing reasonable ranges of possible outcomes for a set of strategic decisions. Bounding results in a narrower range of potential future scenarios to be addressed.

Case: The set of aggregate, macroeconomic market conditions of the indication for which a compound is being evaluated. Information on epidemiology, changing demographics, treatment paradigms, historical market revenue trends, and existing competitors, all define the Case for an NCE considering entry into a new marketplace.

Compliance: A continuous measure of the number of pills a patient actually takes over a given interval, compared to the number prescribed. Compliance affects revenue generation in pharmaceutical marketplaces by effectively increasing the length of the prescription, resulting in fewer prescriptions filled.

Connector: A building block of the Dynamic Modeling language that allows information to be transferred among model variables. Connectors are used to show relationships between individual pieces of model structure, allowing variables to be a function of other conditions within a model.

Continuous models: Dynamic models in which items are grouped into collections of elements. Continuous models may be disaggregated into detailed segments, but each segment still represents a group of items rather than unique and distinct individuals, as in an agent-based model.

Cumulative discounted revenue: A single metric for model analysis whereby revenue streams for a compound are discounted back to the current timeframe using an appropriate discount rate.

Dimension: A broad class along which patients/physicians/compounds will be grouped in the disaggregation of a Dynamic Model.

Element: A grouping choice within a dimension that determines the classifications of disasaggregation. Elements are often chosen based on an MECE concept.

Expected sales trajectory: The most likely forecasted scenario for sales/prescriptions of a newly released compound, sometimes referred to as uptake. This projection is often the result of analysis of historical analogs, but can be the product of Dynamic Modeling as well, and includes the peak volume/sales, the speed at which the compound will reach that mark, and the shape of the falloff from that point.

External influence: The effect of marketing factors on the diffusion of a product, such as the impact of detailing, samples, journal advertising, and medical education on the adoption of pharmaceuticals by physicians.

Feedback: A series of interrelationships whereby a particular variable affects other model components that ultimately "feed back" to further change the original variable. Feedback comes in two types—balancing and reinforcing.

Flow: A building block of the Dynamic Modeling language that moves items from place to place (usually into or out of a stock) over a period of time. Flows are the fundamental drivers of change in any system and are the focus of any potential intervention strategy.

Internal influence: The effect of word-of-mouth factors on the diffusion of a product, such as the impact of doctors encouraging each other to try certain pharmaceutical products.

Longitudinal integrated claims (LIC) databases: Patient-level data sources that cross-reference pharmacy claims with specific disease diagnoses and track individual patient behavior over time. LIC databases are a key component of the data collection process for Dynamic Modeling efforts.

Lines of therapy: A set of sequential treatment choices as a patient advances through various forms of prescribed medication. Pharmaceutical firms sometimes seek lines of therapy designations in their labeling from the FDA, such that a compound may be indicated as a first-line therapy in an indication, for example.

Maximum entropy: An estimation technique that extracts the maximum amount of information about the model parameter values from the assumptions made about the ranges of potential future outcomes.

MECE (mutually exclusive, collectively exhaustive): Often used to define elements of a dimension in a dynamic model, ensuring every entity is captured once and only once in each defined segment.

Mental models: Implicit cause-and-effect relationships that are often the basis for decision making. Mental models are often simulated implicitly when trying to determine the results of an action or strategy.

Monte Carlo analysis: Analytic technique in which a large number of simulations are performed independently and the corresponding results are grouped together. The input assumptions for each simulation are drawn from a distribution, so the outcomes have a corresponding distribution from which confidence bounds for results can be established.

Most likely scenario: The most probable outcome for a given market metric, often the prescription volume or revenue projection for an NCE. Most Likely Scenarios incorporate uncertainty related to important model parameters, often by employing Monte Carlo analysis.

New chemical entity (NCE): A chemical compound developed by a pharmaceutical firm for its effect on the biologic processes involved in one or more diseases. NCEs generally refer to compounds in the development stages or in the very early part of their launch cycle.

New molecular entity (NME): Similar to an NCE, but referring to large biologic molecules that replicate naturally occurring molecular structures in living organisms. NMEs are usually produced by biotechnology firms and are sometimes referred to as new biologic entities (NBCs).

Pace: The speed at which an NCE can reach its long-run place in a disease market. Effective Brand Planning leverages the observed marketplace dynamics to increase the Pace for the compound.

Persistence: A measure of the average time patients spend on a particular therapy, which in fact encompasses both switching and discontinuation dynamics. Market-level persistence refers to the length of time patients spend on any form of prescription treatment before becoming Non-Persistent and not taking any medication.

Place: The position of a new compound in terms of its clinical profile and where it is expected to fall within a defined Lines of Therapy progression. In economic language, the Place of a compound in an indication determines its long-run market share potential—often referred to as peak sales.

Reinforcing feedback: A feedback mechanism characterized by ever-increasing change in a single direction. Reinforcing Feedback produces virtuous or vicious cycles that run out of control unless checked by some sort of Balancing Feedback mechanism.

Robust decision making: A technique for analyzing situations involving deep uncertainty using regret as a measure of robustness. A set of decisions is said to be more robust if it has lower regret than alternative strategies.

Segment: A disaggregation option where interelement movement is either theoretically impossible, possible but not easily quantifiable, or beyond the scope or intent of the modeling effort. If elements are defined as segments, individuals within an element designation never change that categorization.

Steady state: A condition in which the values of the stocks in a system are not changing significantly over time.

Stock: A building block of the Dynamic Modeling language that houses where items exist at a single time point. Stocks represent accumulations of items that results from a net imbalance of the flow(s) associated with them.

Strategic sensitivity grid: A two-dimensional matrix on which the results of sensitivity analysis may be placed in order to assess the tradeoffs between achievability and impact of various strategic options. Strategic Sensitivity Grids help prioritize a set of potential strategies by identifying those that have the lowest cost for a given level of effectiveness.

Structure/behavior paradigm: A central tenet of the Dynamic Modeling approach that says that the observed behavior of any system is a function of the operational structure that drives it. The behavior can be improved and/or changed only by understanding and modifying the structure in order to achieve a set of desired results.

Index